MOVING LOS ANGELES

Short-Term Policy Options for Improving Transportation

AF701723

Paul Sorensen • Martin Wachs • Endy Y. Min • Aaron Kofner • Liisa Ecola
Mark Hanson • Allison Yoh • Thomas Light • James Griffin

Sponsored by James A. Thomas, the L.A. County Metropolitan Transportation Authority, the Music Center of Los Angeles County, and the RAND Corporation

RAND Transportation, Space, and Technology

A RAND INFRASTRUCTURE, SAFETY, AND ENVIRONMENT PROGRAM

This study was sponsored by James A. Thomas, the L.A. County Metropolitan Transportation Authority, the Music Center of Los Angeles County, and the RAND Corporation and was conducted under the auspices of the Transportation, Space, and Technology (TST) Program within RAND Infrastructure, Safety, and Environment (ISE).

Library of Congress Cataloging-in-Publication Data

Moving Los Angeles : short-term policy options for improving transportation/ Paul Sorensen ... [et al.].
p. cm.
Includes bibliographical references.
ISBN 978-0-8330-4555-3 (pbk. : alk. paper)
1. Traffic congestion—California—Los Angeles. I. Sorensen, Paul, 1967– II. Title.

HE356.5.L7S56 2008
388.3'142—dc22

2008035001

The RAND Corporation is a nonprofit research organization providing objective analysis and effective solutions that address the challenges facing the public and private sectors around the world. RAND's publications do not necessarily reflect the opinions of its research clients and sponsors.

RAND® is a registered trademark.

Cover photo © 2008 Jupiterimages Corporation

© Copyright 2008 RAND Corporation

All rights reserved. No part of this book may be reproduced in any form by any electronic or mechanical means (including photocopying, recording, or information storage and retrieval) without permission in writing from RAND.

Published 2008 by the RAND Corporation
1776 Main Street, P.O. Box 2138, Santa Monica, CA 90407-2138
1200 South Hayes Street, Arlington, VA 22202-5050
4570 Fifth Avenue, Suite 600, Pittsburgh, PA 15213-2665
RAND URL: http://www.rand.org/
To order RAND documents or to obtain additional information, contact
Distribution Services: Telephone: (310) 451-7002;
Fax: (310) 451-6915; Email: order@rand.org

Preface

The Los Angeles (L.A.) region, according to many studies, has the most severe traffic congestion in the United States. Longer-term trends in many of the underlying causal factors—including growth in the population, the economy, and the movement of goods through the Ports of Los Angeles and Long Beach—suggest that congestion will continue to worsen in the coming years, absent significant policy intervention. Excessive traffic congestion detracts from quality of life, is economically wasteful, is environmentally damaging, and exacerbates social-justice concerns. Finding efficient and equitable strategies for mitigating congestion will therefore serve many social goals.

The primary intent of this study was to recommend strategies for reducing congestion in L.A. County that could be implemented and produce significant improvements within a short time, defined as roughly five years or less. Specific elements of the study include the following:

- reviewing the academic literature for insights on congestion
- examining available data to characterize traffic congestion in Los Angeles, including current conditions and recent trends
- examining relevant transportation and land-use features in Los Angeles to diagnose the key contributors to congestion in the region
- identifying the range of available congestion-reduction strategies that could be implemented and produce effects within the near term

- assessing the strengths and weaknesses of each option with regard to cost/revenue implications, short- and longer-term effectiveness in reducing congestion, effects on other social goals, likely implementation obstacles, and the current level of implementation in Los Angeles
- recommending a smaller set of strategies that offers the greatest prospects for reducing congestion and improving transportation options in Los Angeles
- considering complementary strategies for building political consensus around effective, albeit potentially controversial, congestion-reduction measures.

The study was sponsored by a small consortium of public and private donors sharing an interest in reducing traffic congestion through improved transportation policy in Los Angeles, including James A. Thomas, the L.A. County Metropolitan Transportation Authority, the Music Center of Los Angeles County, and the RAND Corporation.

The intended audience includes community leaders and elected officials in L.A. County along with other interested residents in the region. Though the specific recommendations proffered in the book are tailored to the L.A. region, leaders in other cities who are interested in strategies to reduce congestion should also find the underlying analysis to be of value.

The RAND Transportation, Space, and Technology Program

This research was conducted under the auspices of the Transportation, Space, and Technology (TST) Program within RAND Infrastructure, Safety, and Environment (ISE). The mission of ISE is to improve the development, operation, use, and protection of society's essential physical assets and natural resources and to enhance the related social assets of safety and security of individuals in transit and in their workplaces and communities. The TST research portfolio encompasses policy areas including transportation systems, space exploration, information and telecommunication technologies, nano- and biotechnologies, and other aspects of science and technology policy.

Questions or comments about this monograph should be sent to the project leader, Paul Sorensen (Paul_Sorensen@rand.org). Information about the Transportation, Space, and Technology Program is available online (http://www.rand.org/ise/tech/). Inquiries about TST research should be sent to the following address:

Martin Wachs, Director
Transportation, Space, and Technology Program, ISE
RAND Corporation
1776 Main Street
P. O. Box 2138
Santa Monica, CA 90401-2138
310-393-0411, x7720
Martin_Wachs@rand.org

Contents

Figures

Tables

Abbreviations

AB	assembly bill
BRT	bus rapid transit
DOT	department of transportation
FSP	Freeway Service Patrol
GPS	Global Positioning System
HOT	high-occupancy toll
L.A.	Los Angeles
LACDPW	Los Angeles County Department of Public Works
LADOT	City of Los Angeles Department of Transportation
LAX	Los Angeles International Airport
Metro	Los Angeles County Metropolitan Transportation Authority
mph	miles per hour
PeMS	Freeway Performance Measurement System
RTPA	regional transportation planning authority
SOV	single-occupant vehicle
TDM	transportation demand management

TSM	transportation system management
TST	Transportation, Space, and Technology
TTI	Texas Transportation Institute
V/C	volume to capacity
VMT	vehicle-mile traveled

Summary

The time is right for aggressive transportation-policy reform in Los Angeles. Though rising fuel prices have reduced traffic on some routes in recent months, the broader trend is that traffic conditions on the region's highway and road network have steadily deteriorated over a period of many years. Frustrations with congestion remain high, and many residents and policymakers agree that "something must be done!" Traffic congestion is not merely annoying—it lowers quality of life, detracts from economic competitiveness, leads to additional traffic accidents, wastes fuel, produces excess greenhouse-gas emissions, exacerbates air-quality problems for communities in the vicinity of crowded freeways, and slows bus service for those who rely on transit. The buzz in print and on the Web implies an increased willingness to consider more innovative policies for solving the region's congestion problem.

Past debates about reducing congestion often revolved around how best to spend limited transportation dollars. Should the region invest in more buses or rail lines, add more left-turn lanes and signals, augment the traffic-signal timing and control system to reduce intersection delays, or widen freeway bottlenecks? While such investments may slow the rate at which congestion gets worse, they are fundamentally unable to halt or reverse the growth of congestion in the face of increasing automotive travel. Neither do the most commonly proposed land-use strategies offer much hope on their own.

A key finding from the RAND Corporation study described in this book is that strategies that rely on pricing to manage the *demand* for driving—e.g., by charging more for driving and parking during peak hours in the most congested locations—are extremely effective in producing sustainable reductions in congestion. Pricing strategies lead to more-efficient use of existing road capacity and can raise substantial revenues to fund needed transportation improvements. Though certain forms of pricing may lead to concerns about the ability of lower-income drivers to pay the charges, the research suggests that investing the resulting revenue in much-improved transit services and other nonautomotive travel options can be effective in mitigating such concerns.

Policies that improve alternatives to driving alone (such as transit, carpooling, biking, or walking) can play an important role in improving transportation in Los Angeles, as can supply-management strategies designed to enhance the capacity and efficiency of existing streets and freeways. But one of the keys to long-term congestion relief will be an integrated package of reforms, and efforts to promote transportation alternatives and improve the existing road network will be much more effective if implemented in concert with strategies for managing the demand for peak-hour automotive travel through the proven mechanism of pricing. In fact, the research clearly demonstrates that any package of reforms that does not include pricing strategies will not achieve lasting reductions in traffic congestion. As a region, then, L.A.-area stakeholders must summon the political willpower to face a tough decision. Will Los Angeles begin to pursue pricing to manage demand for peak-hour automotive travel, or will it instead simply allow congestion to worsen in the coming decades? These are the only choices.

Strategies for Reducing Congestion in L.A. County and Recommendations That Can Be Implemented and Produce Results Quickly

The principal goal of the RAND study was to develop short-term congestion-reduction recommendations for application within L.A. County—that is, strategies that could be implemented and lead to marked reductions in peak-hour traffic delays within a period of approximately five years. We paid specific attention to strategies that would prove helpful in dense urban areas, where congestion is often the most intense, though some of the strategies considered should also be useful in suburban locales. We constrained the focus to policy options applicable to passenger traffic; while some of the strategies might apply to truck traffic as well, we did not include strategies specifically targeted at goods movement.

These criteria, adopted to establish a logical and reasonable bound on the study's scope, precluded including certain potentially promising strategies that might be pursued in Los Angeles. The short-term nature of the recommendations, for instance, ruled out major capacity expansions, such as new freeway lanes or rail transit lines, which take far longer to plan, approve, fund, and build. It also removed from consideration major land-use policies related to zoning, parking, urban density, and transit-oriented development. Though such reforms could be implemented within five years, their effects would unfold much more slowly—over a period of decades—as land-use patterns evolve with successive waves of redevelopment. As another example, given the decision to focus on passenger- rather than freight-oriented strategies, we did not consider the provision of grade separations at rail crossings, though it could offer significant benefits in some parts of L.A. County.

Even with these scope limitations, however, there are still many viable options for reducing traffic congestion in the near term. Broadly, these include strategies for making more-efficient use of existing roadways, strategies for managing demand for peak-hour driving, and strategies for improving bus transit and other nonautomotive travel modes.

During the course of our study, we reviewed findings from the literature, examined available data, and spoke with agency staff and elected officials in the county. Our efforts were aimed at the following:

- understanding congestion at the conceptual level
- characterizing congestion in L.A. County
- diagnosing the severity of congestion in L.A. County
- identifying and evaluating short-term congestion-reduction strategies
- developing short-term congestion-reduction recommendations

- developing complementary recommendations to help overcome political obstacles and build the consensus necessary for implementation.

This document summarizes the study's findings related to congestion in general and to the most promising congestion-reduction strategies for jurisdictions operating within L.A. County in particular. Because congestion plagues many other metropolitan areas in the United States, the discussion in this book should be of interest to readers in other urban locales as well. The prime focus, however, is on Los Angeles (as a point of clarification, note that we often use the generic term *Los Angeles* in describing conditions applicable to the region as a whole; where greater jurisdictional specificity is required, we explicitly refer to either L.A. County or the City of Los Angeles).

The results of our analysis led to 13 integrated short-term recommendations for reducing traffic congestion on the network of arterial roads and highways in L.A. County. The recommendations feature several pricing strategies for managing peak-hour automotive travel and raising transportation revenue, along with complementary strategies for improving alternative transportation options and boosting the efficiency of the existing road network.

Among the recommendations, those involving the use of pricing are likely to stir the greatest debate. The book cites numerous real-world examples illustrating the success of existing pricing programs and outlines compelling theoretical arguments to show that pricing—by requiring that drivers consider the full social and environmental costs in their travel decisions—helps to promote greater social and economic welfare. Even so, there will be those who view themselves as worse off under a system of pricing, and there will also be those who oppose the idea on principle. With this in mind, we chose to consider complementary strategies for mitigating concerns and building political consensus as well.

Historically, in Los Angeles, when circumstances have indicated both the need and the opportunity for change, the business community has responded by galvanizing support for significant transportation reforms. Successful action on the congestion problem will now require the collaboration of a much broader group of interested parties who can initiate and sustain forward momentum and ensure that the concerns of all affected groups are considered. Though the transportation-policy recommendations developed in this book would be implemented by public agencies, the complementary strategies for building political support are intended to assist community leaders who would like to play a role in promoting meaningful efforts to reduce traffic congestion and improve transportation options in Los Angeles.

What Do We Know About Congestion?

In considering congestion-reduction strategies for Los Angeles, we found it helpful to first take stock of accumulated wisdom based on the prior work of transportation researchers and scholars. The following points highlight several key insights.

Congestion Results from an Imbalance Between the Supply of Road Capacity and the Demand for Driving During Peak Travel Hours

Potential solutions thus include managing peak-hour demand and boosting supply. Until supply and demand are brought into closer alignment, however, congestion will resolve the imbalance by making drivers wait their turn to use the available road capacity.

Growth in Automotive Travel Has Far Exceeded Road Expansion in Recent Decades

One of the reasons that demand outstrips supply is that the number of vehicle-miles traveled (VMT, a common measure of automotive travel) has been growing much more quickly than has the nation's road supply for many years now. This is shown in Figure S.1, which compares growth in road lane–miles, population, the economy (as measured by gross domestic product, or GDP), and VMT in the United States since 1970. During this period, the supply of lane-miles has been relatively stagnant, while growth in VMT has far exceeded growth in the population and, in fact, tracks quite closely with GDP. Assuming continuation in these underlying trends, the gap between VMT and

Figure S.1
Growth in Lane-Miles, Population, Gross Domestic Product, and Vehicle-Miles Traveled in the United States Since 1970

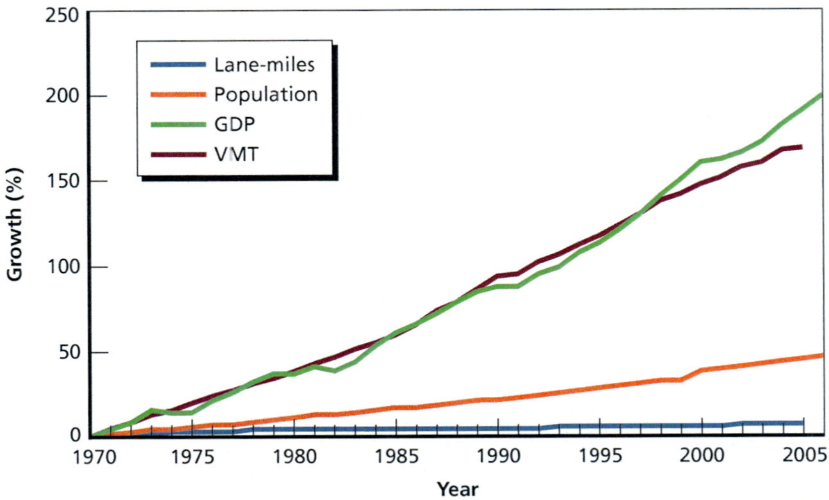

SOURCES: California Department of Industrial Relations (2007); Caltrans (undated, 2008).
RAND *MG748-S.1*

road supply is likely to further widen in the coming years, leading to even more traffic congestion.

Transportation-Revenue Shortfalls Preclude Los Angeles from Building Its Way Out of Congestion

Federal and state motor-fuel excise taxes, responsible for a significant share of transportation funding, are levied on a cents-per-gallon basis and thus need to be raised periodically to offset the effects of inflation and improved fuel economy. Wary of antitax sentiments, legislators have become increasingly reluctant in recent years to address this politically unpopular task; in California, the gas tax was last raised in 1994, while the last increase in the federal gas tax occurred in 1993. As a result, Los Angeles now collects far less real revenue per mile of vehicle travel than in years past. Consider Figure S.2, which shows growth in the nominal per-gallon excise gas tax in California compared with growth in inflation and the cost of construction since 1972.

Figure S.2
Growth in California Gas Tax, General Inflation, and Highway-Construction Costs

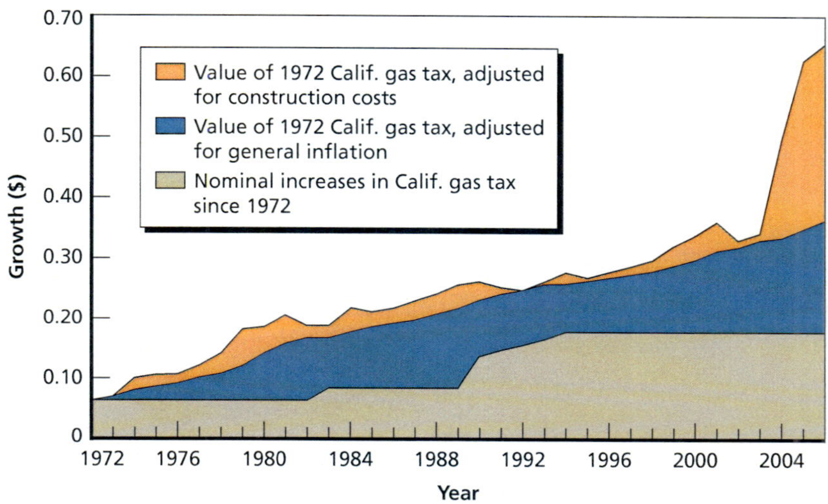

SOURCES: California Department of Industrial Relations (2007); Caltrans (undated, 2008).
RAND *MG748-S.2*

While the gas tax has been raised several times and now stands at $0.18 per gallon, during the same period, the value of the $0.07 gas tax in 1972 has risen to $0.36 with inflation and to $0.65 in terms of construction costs. Just to keep pace with inflation, then, the current California excise gas tax would need to be doubled; to keep pace with highway-construction costs, it would need to be more than tripled. In short, the gas tax no longer buys what it used to. Absent a significant boost in available transportation revenue, Los Angeles's ability to provide new road (or transit) infrastructure is severely limited.

Failure to Charge the Full Costs Associated with Automotive Travel Inflates the Demand for Driving

From an economic perspective, automotive travel is underpriced. Driving a vehicle creates environmental and social costs, such as harmful emissions and additional congestion delays for other travelers. When we choose to drive, we are not forced to pay for these so-called external costs; rather, they are passed along to other members of society, such as drivers traveling in our traffic wake or residents living alongside the freeways on which we travel. Because driving is underpriced, society tends to overconsume road space; that is, we make many trips for which the total costs (including external costs passed on to others) exceed the total benefits. In theoretical terms, this reduces social welfare. In practical terms, it leads to greater traffic congestion and contributes to environmental problems, such as poor air quality and additional greenhouse-gas emissions.

Small Changes in Driving Can Lead to Large Changes in Congestion

The relationship between the number of vehicles and their travel speed can be described as *nonlinear.* When there are just a few cars on the road, more can be added without having much effect on travel speed. When the road is already crowded, on the other hand, adding just a few more cars can trigger congestion, significantly reducing travel speed and simultaneously diminishing the road's effective capacity (that is, reducing the number of vehicles per lane per hour that the road can carry). The practical import of this observation is that when a road is already congested, reducing the number of cars trying to use that road at the same time by even a small amount can often produce much larger reductions in congestion delays. For example, reducing the number of cars on the road by 2 or 3 percent might cut congestion delays by 10 or 15 percent.

The Easy Solutions to Congestion Have Already Been Implemented

Congestion is not a new phenomenon, and the easy solutions—those that are effective, inexpensive, and uncontroversial—have long since been applied. Remaining options tend to be costly, controversial, or only moderately effective. This leaves Los Angeles with two options. It can either continue to implement a wide range of relatively inexpensive and uncontroversial measures that offer modest benefits in the hope that their combined effects will be significant, or it can strive to overcome financial and political obstacles and pursue the options that appear to hold the greatest promise.

Few Congestion-Reduction Strategies Remain Effective in the Longer Term

A phenomenon described as *triple convergence* (Downs, 2004) undermines the effectiveness of many congestion-reduction strategies. In short, when traffic conditions on a roadway are improved during peak hours, additional travelers will tend to converge on that newly freed capacity from (1) other times of travel, (2) other routes of travel, or (3) other modes of travel, slowly eroding the initial peak-hour congestion-reduction benefits in the busiest travel corridors. Longer-term increases in the demand for automotive travel resulting from population growth and economic expansion can further undermine a strategy's effectiveness. This is why we often see, for instance, that flow improves for a short while when new lanes are added to a freeway but usually returns to former levels of congestion within just a few years. This is not to suggest that such improvements lack merit; they may, for instance, afford a greater level of aggregate travel, improve travel choices, or reduce the spatial and temporal spread of congestion. Rather, the key point is that they will prove unable to reduce peak-hour congestion in the busiest corridors for more than a short period.

Only Pricing Strategies Can Produce Sustainable Reductions in Traffic Congestion

The only strategies resistant to the effects of triple convergence involve the use of pricing to manage the demand for peak-hour automotive travel (Downs, 2004). Often described as *congestion pricing*, examples include charging higher tolls to drive during peak hours or charging higher prices to park in the most convenient curb spaces at the busiest times of day. The main reason that the effectiveness of pricing strategies is not eroded by triple convergence is that the same peak-hour charges that encourage some to change their travel patterns also deter others from converging on the freed capacity. Another way of stating this is that pricing strategies represent the only approach that can reduce congestion without inducing additional automotive-travel demand. Pricing also remains effective in the longer term in the face of generally increasing demand, provided that the prices charged are allowed to rise with demand. Pricing strategies can help raise needed revenue as well, and, by preventing congestion, they facilitate more-efficient use of existing capacity.

Is Congestion in Los Angeles Really So Bad?

Angelenos complain about traffic congestion, and rightly so. Some overseas cities, such as Bangkok, Jakarta, and Lagos, are more congested, but, by most measures, traffic conditions in Los Angeles are indeed worse than in any other major U.S. metropolitan area.

Since beginning this study in the summer of 2007, the economy has weakened and fuel prices have surged. Drivers have looked for ways to reduce their travel, and data from California's Freeway Performance Measurement System (PeMS) suggest that traffic delays have declined on some local freeways in the past six to nine months. Yet the economy will, at some point, recover, and consumers will increasingly switch to more fuel-efficient vehicles should gas prices remain high. As the population continues to grow and as goods continue to flow through the Ports of Los Angeles and Long Beach, traffic congestion will sooner or later begin to worsen once again.

This is a problem, as congestion in Los Angeles (even with recent declines) is already quite severe. According to the Texas Transportation Institute (TTI), which develops widely cited congestion statistics for large U.S. metropolitan areas (Schrank and Lomax, 2007), the greater L.A. metropolitan area (as defined by the U.S. Census Bureau) consistently leads the nation in the following:

- total annual hours of delay for all travelers (490 million)
- total annual gallons of wasted fuel for all travelers (384 million)
- average annual hours of delay per peak-period traveler (72 hours)
- average annual gallons of wasted fuel per peak-period traveler (57)
- total annual economic costs as a consequence of congestion delay (more than $9 billion, up from $2 billion in 1982).

More-detailed analyses of the freeway and arterial (street) networks in L.A. County reveal that congested travel conditions are ubiquitous on the freeway network, that freeway-travel times are extremely unreliable from one day to the next, that arterial congestion is especially intense on the Westside, and that truck traffic is most severe on the highways and around the ports and downtown Los Angeles.

Illustrating the first point, the map in Figure S.3 shows the average number of hours per day, for weekdays in 2006, during which the travel speed on different links in the L.A. County freeway network averaged less than 35 miles per hour (mph) (averaged across both directions of flow). Congestion appears to be especially severe on Interstate 5 near the Orange County border and on U.S. Route 101 (the 101

Figure S.3
Hours per Day During Which Freeway Speeds Average Less Than 35 mph in General-Purpose Lanes, 2006 Weekdays

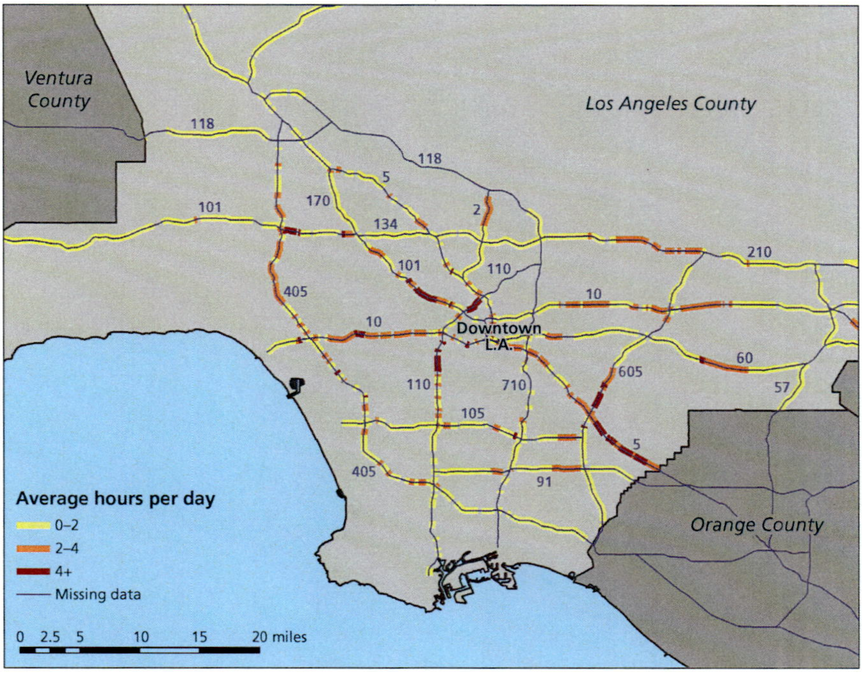

SOURCE: PeMS (undated).
RAND *MG748-S.3*

freeway) just northwest of downtown Los Angeles. Generally, though, congestion is prevalent throughout much of the freeway network, with many links experiencing four or more hours per day of travel speeds averaging less than 35 mph. And weekend travel conditions are not much better.

Figure S.4 captures the pattern of congestion on the arterial-street network in Los Angeles. Specifically, the map shows estimates of the volume-to-capacity (V/C) ratio for different arterial links in the afternoon peak travel hours as of 2004 (as the V/C ratio approaches 1.0, shown in darker lines on the map, congestion intensifies). Here we see that, while there are many congested arterials throughout the county,

Figure S.4
Modeled Arterial Volume-to-Capacity Estimates for the Afternoon Peak, 2004

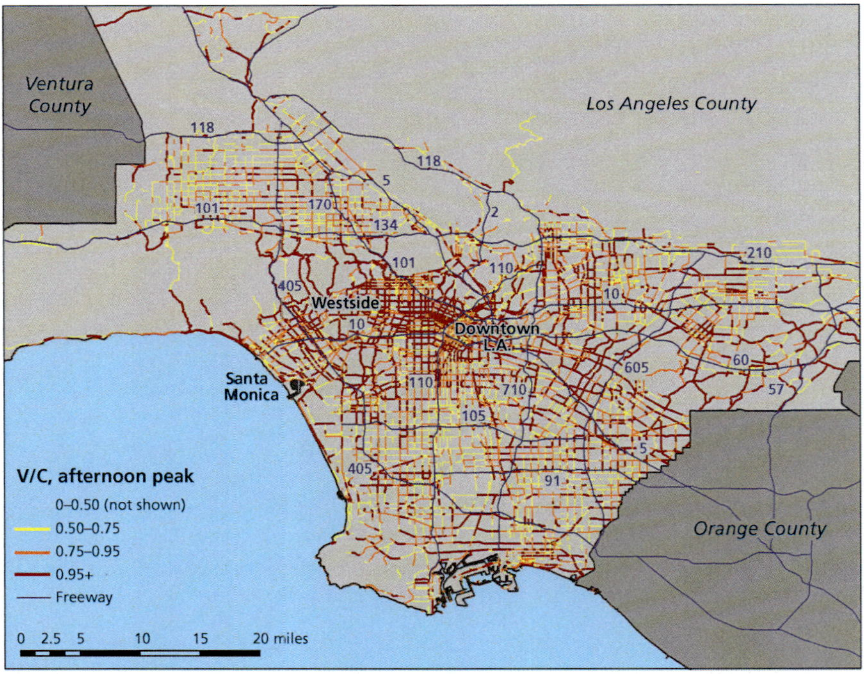

SOURCE: 2004 regional transportation-model data provided by SCAG staff.
RAND *MG748-S.4*

the pattern is especially pronounced between downtown Los Angeles and the Westside.

What Makes Los Angeles Different?

To further inform the development of suitable congestion-reduction strategies for Los Angeles, we took a closer look at some of the underlying factors that contribute to the region's congestion. What is it about Los Angeles, specifically, that leads to the most severe congestion in the nation, and what implications does this have for the types of strategies that might offer the greatest prospects for reducing congestion?

Simple Explanations Fail to Explain the Severity of Congestion in Los Angeles

The results of our inquiry offered some surprises. Many of the assumptions that observers might make about the causes of congestion in Los Angeles turn out to be either incomplete or inaccurate.

L.A. residents do not drive more than other urban dwellers. There is a common perception that Southern Californians have a longstanding love affair with their automobiles, and the severity of congestion in Los Angeles reinforces the assumption that residents of the region drive more than their counterparts elsewhere. In fact, this turns out not to be the case. Among the 14 largest metropolitan areas evaluated in TTI's annual mobility studies (Schrank and Lomax, 2007), the greater L.A. region ranks just

- fifth in daily per capita VMT, after Dallas, Houston, Atlanta, and Detroit
- fifth in average household automobile ownership, after Seattle, Atlanta, San Francisco, and Dallas
- ninth in the percentage of employees who drive to work alone, after Detroit, Dallas, Houston, Atlanta, Miami, Phoenix, Boston, and Philadelphia.

Los Angeles has a very extensive and well-managed road network. Another possible explanation might be that the region simply needs to add more road capacity. Yet among the 14 largest metropolitan areas considered by TTI, Los Angeles has by far the densest road network, providing more than 50 percent more lane-miles per square mile than Detroit, its nearest competitor. Even when framed in terms of lane-miles per capita, Los Angeles still ranks eighth among the 14 largest metropolitan areas. Moreover, state and local transportation agencies have implemented sophisticated programs, such as ramp metering and synchronized traffic signals, to use the road system as efficiently as possible.

Los Angeles provides a significant level of transit service. Proponents of alternative transportation might argue that the severity of congestion in Los Angeles stems not from a lack of sufficient road supply, but rather from inadequate provision of competitive transit services. Yet according to statistics from the American Public Transportation Association (APTA, 2007) and TTI (Schrank and Lomax, 2007), the transit system in Los Angeles appears robust in comparison to many other urban areas. Of the 14 largest metropolitan areas evaluated by TTI, Los Angeles ranks

- second in total bus-service miles, after New York
- first in bus-service miles per square mile
- third in bus-service miles per capita, after San Francisco and Washington, D.C.
- fifth in total rail-transit track-miles (including commuter rail, light rail, and subways), after New York, Chicago, Philadelphia, and Boston
- seventh in rail-transit track-miles per square mile, after New York, Chicago, Philadelphia, San Francisco, Boston, and Washington, D.C.
- seventh in rail-transit track-miles per capita, after Philadelphia, Boston, New York, Chicago, San Francisco, and Washington, D.C.

High Regional Population Density Is a Key Contributor to Congestion in Los Angeles

Despite its reputation for sprawling development, Los Angeles is quite densely populated at the regional scale. While downtown Los Angeles is not as dense as, say, Manhattan or downtown Chicago, the suburbs surrounding Los Angeles are much denser than the suburbs surrounding other major cities (Manville and Shoup, 2005). As a result, Los Angeles is by far the densest metropolitan area in the country.

As density increases, individuals tend to drive less on a per capita basis. This is because trip origins and destinations are often closer together, leading to shorter car trips, and people can rely on alternatives, such as walking, biking, or transit, for a larger share of trips. Yet this effect can be overwhelmed by the fact that there are also more drivers competing for the same road space, thus intensifying traffic congestion (Manville and Shoup, 2005). For instance, though Los Angeles has by far the densest road network among major metropolitan areas in the United States, as already noted, it still ranks second in terms of total VMT per total lane-miles, just behind San Francisco (based on data from Schrank and Lomax, 2007). In short, greater population density tends to exacerbate congestion, and Los Angeles is very dense.

High population density can also combine with other factors to make congestion worse. We mentioned earlier that L.A. residents do not drive more than residents of other large areas. It turns out, however, that they drive a lot on a per capita basis considering the region's density; in other words, Angelenos do not seem to curtail their driving as much as one might expect in response to higher density. Figure S.5 compares regional population density with daily per capita VMT for the country's largest 14 metropolitan areas.

For most of the cities shown in the figure, there is a fairly consistent relationship in which per capita VMT declines with regional density. Los Angeles stands as an exception. The only other large metropolitan areas in the country with higher per capita VMT (Atlanta, Dallas, Houston, and Detroit) are all much less dense than Los Angeles. For regions in which the level of density approaches that of Los Angeles, such as San Francisco, Washington, D.C., and New York, per capita VMT is much lower. We thus see a confluence of three density-

Figure S.5
Population Density and Daily Per Capita Vehicle-Miles Traveled in Major Metropolitan Areas

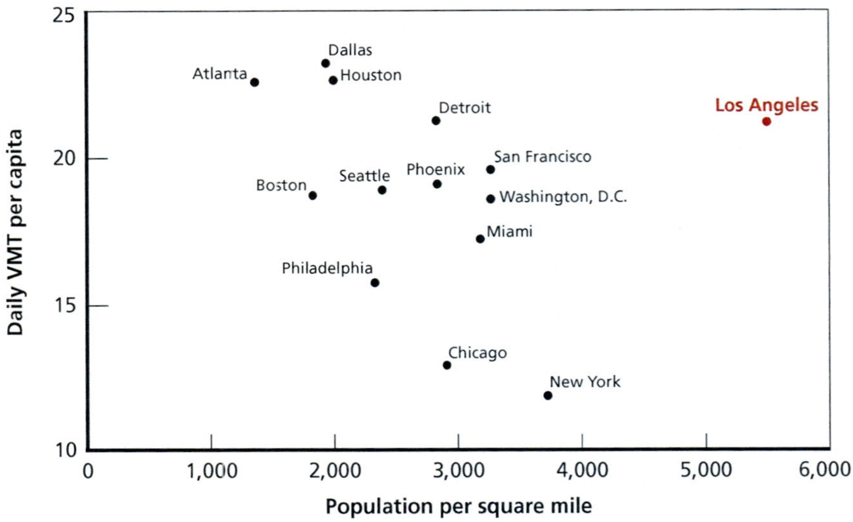

SOURCE: Schrank and Lomax (2007).
RAND *MG748-S.5*

related factors that, in combination, help to explain the severity of congestion in Los Angeles: (1) congestion is likely to rise with increased population density, (2) Los Angeles is much denser than its peers at the regional level, and (3) Los Angeles exhibits a surprisingly high level of per capita VMT relative to its density.

Land-Use Patterns in Los Angeles Make It Harder to Provide Effective Transit and Reinforce Reliance on the Automobile

One of the reasons that L.A. residents may drive more than one might expect given the region's density pertains to the prevailing land-use patterns. Rather than a single dominant downtown area such as one might find in New York or Chicago, Los Angeles has numerous high-density clusters scattered throughout the region—such as Santa Monica, Century City, Long Beach, Glendale, and Pasadena. This pattern, often described as *polycentricity* (multiple centers), can, in fact, ease traffic congestion by spreading car trips out over a greater percentage of the road network's lane-miles. Within the context of Los Angeles, however, polycentricity may serve to exacerbate congestion in two ways.

First, polycentricity makes it more challenging to develop a well-connected, high-speed transit network with dedicated right-of-way to attract more riders and, in turn, reduce automotive travel. To begin with, the network will require more links to connect all of the dispersed population clusters and job centers with one another. Consider, for instance, that Los Angeles has constructed significant light-rail and subway track mileage in the past several decades, yet there are still obvious gaps in the network's coverage, such as between downtown Los Angeles and the Westside and between the San Fernando Valley and the Westside. In addition, the fact that population and jobs are spread out across more centers increases the difficulty of attracting sufficient ridership on any given link to justify the significant investment required for transit lines with dedicated right-of-way.

Second, polycentricity increases the likelihood that residents will need to visit multiple locations to accomplish multiple errands as opposed to taking care of multiple errands all at a single destination. The need to visit multiple locations makes transit—often slower and less convenient than driving—even less attractive by comparison. In short, the polycentricity of Los Angeles makes transit more difficult to provide and less attractive to use, despite the region's high population density, thereby reinforcing automobile dependency—and, in turn, exacerbating congestion.

Cheap and Abundant Parking in Los Angeles Encourages Additional Driving, Further Compounding Congestion

Compared to other large metropolitan areas, and especially in relation to the density of the region, Los Angeles offers abundant and inexpensive parking, and this encourages more people to drive (Manville and Shoup, 2005). In many areas, such as San Francisco, a deliberate effort by planners to reduce private vehicle use limits the number of parking spaces that may be included in a new development. In contrast, developers in most L.A. jurisdictions are required through zoning provisions to provide some minimum number of parking spaces (based on land-use type and project scale), thus ensuring that parking will remain cheap and abundant and cars will remain the dominant mode of travel (Shoup, 2005).

Significant Freight Traffic Also Contributes to the Severity of Traffic Congestion in Los Angeles

Though this book does not consider traffic-reduction strategies directly related to goods movement, in diagnosing the sources of traffic congestion in Los Angeles, it is necessary to acknowledge the significant and growing role of freight traffic in the region. The twin Ports of Los Angeles and Long Beach collectively handle about 43 percent of the container shipments entering the United States, and the volume is expected to increase considerably in the coming decades (SCAG, 2005). Trucks transport many of the containers moving through these ports to inland freight hubs or other regional destinations. While truck traffic is not distributed uniformly throughout the region's road network, it does constitute a significant share of traffic in certain areas and along certain corridors.

What Short-Term Congestion-Reduction Strategies Are Available, and What Are Their Relative Strengths and Weaknesses?

To identify potential short-term measures for reducing congestion in L.A. County, RAND researchers reviewed the scholarly literature, examined prior studies and proposals for Los Angeles, and investigated current approaches being pursued in other large cities in the United States and abroad. This effort led us to consider 28 strategies that can be divided among three broad categories:

- *Transportation system management (TSM).* Strategies in this category, such as traffic-signal timing and control and freeway-ramp metering, are intended to increase the efficiency of the existing road network.
- *Transportation demand management (TDM).* Strategies in this category, such as ride-sharing, flexible work hours, and congestion

pricing, are intended to reduce or manage the demand for automotive travel, especially during peak hours. Subcategories include voluntary or incentive-based programs, regulatory approaches, and pricing strategies.
- *Alternative transportation options.* Strategies in this category, such as bus rapid transit (BRT) and bicycle-infrastructure improvements, are intended to increase the attractiveness or lower the price of alternatives to driving. Subcategories include transit strategies and nonmotorized strategies (bicycle and pedestrian improvements).

The full set of the 28 strategies is shown in Figure S.6. To gain insight into their relative advantages and limitations, we evaluated each option against a broad range of criteria:

- net cost/revenue implications for local government
- short-term effectiveness in reducing congestion
- longer-term effectiveness in reducing congestion
- accessibility, mobility, and traveler choice
- safety
- economic efficiency
- environment
- equity
- interest-group concerns
- general political obstacles
- institutional or jurisdictional challenges
- level of current implementation in Los Angeles.

We considered available options at the strategic level rather than looking at project-specific details, and the scope of the analysis did not allow for formal transportation-system modeling. We thus found it necessary to develop a qualitative system of ratings for characterizing each strategy with respect to these criteria (for example, a strategy might receive a rating of *negligible*, *low*, *medium*, or *high* to describe its longer-term effectiveness in reducing congestion). The strategy ratings were based on available evidence from the research literature (for

Figure S.6
Qualitative Ratings for the 28 Strategies Considered

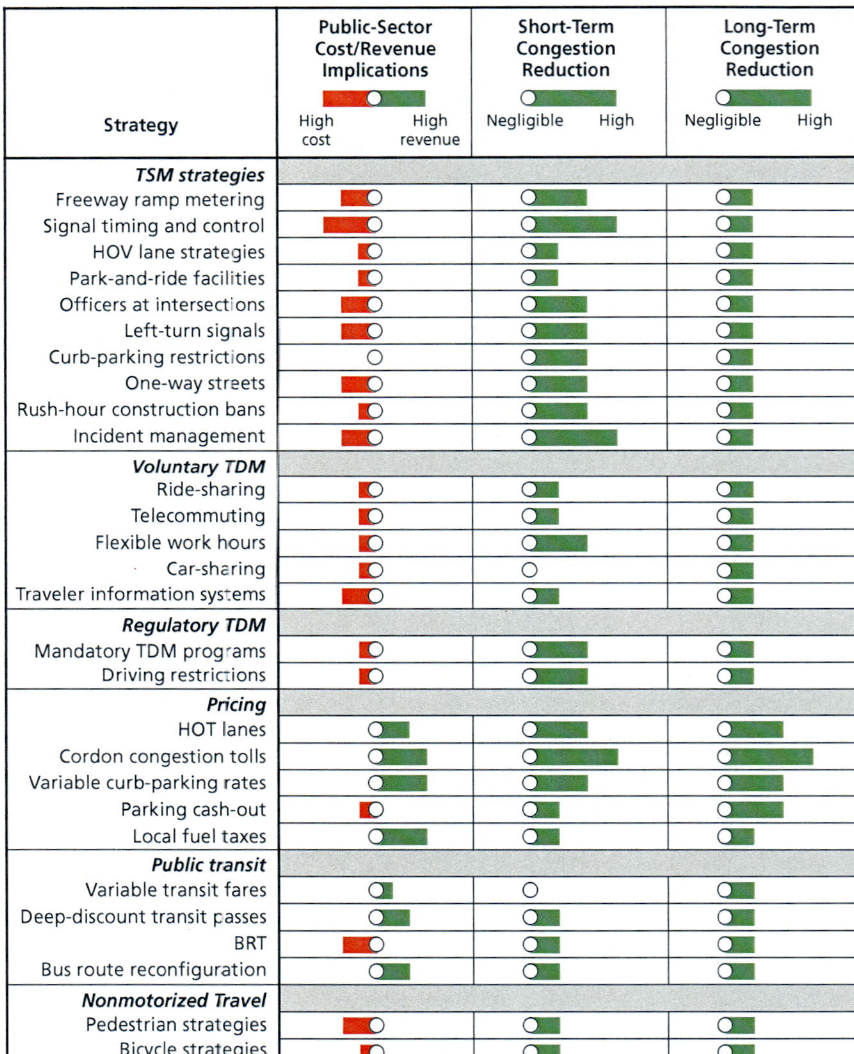

RAND *MG748-S.6a*

performance-related criteria) as well as interviews with knowledgeable staff at state, regional, and local transportation agencies operating in L.A. County (for the current level of implementation in Los Angeles).

Figure S.6—Continued

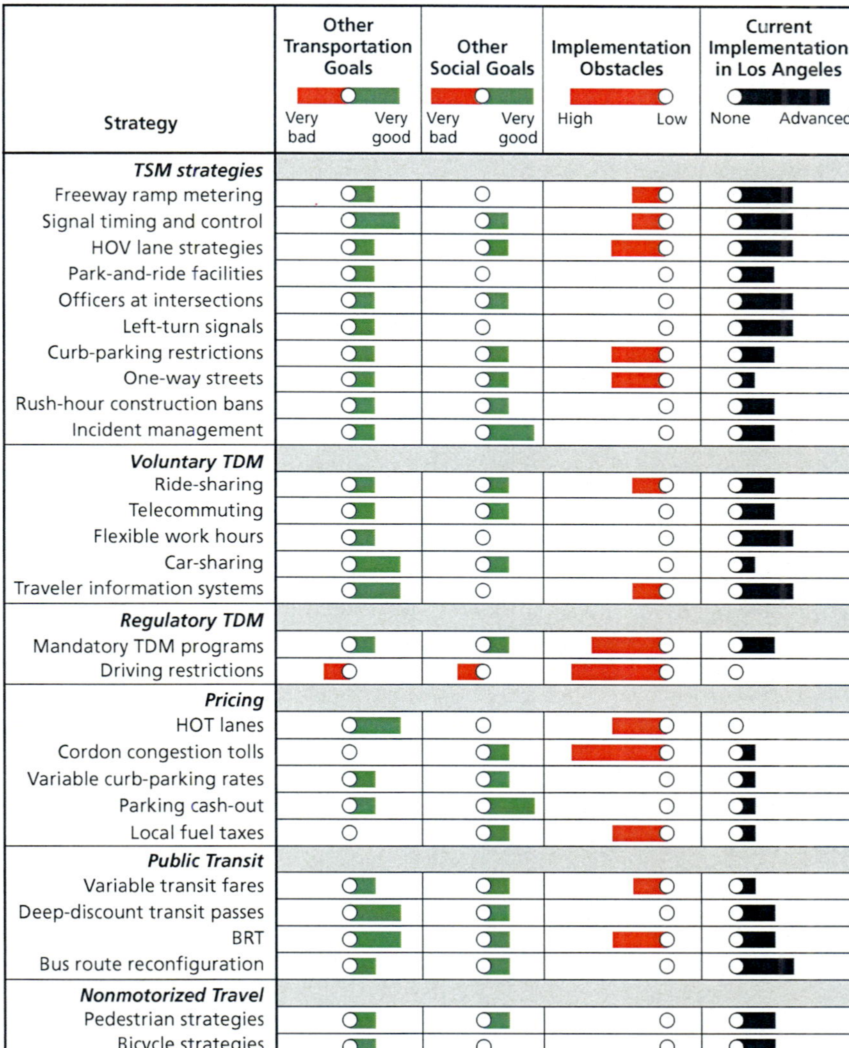

NOTE: HOT = high-occupancy toll.

RAND *MG748-S.6b*

In judging the potential effects of a strategy, we assumed that implementation efforts would be well planned and systematically applied throughout the county in sensible contexts.

Figure S.6 lists the 28 strategies evaluated and summarizes the qualitative ratings for all of the criteria considered. Note that the "Other Transportation Goals" column refers to accessibility, mobility, and traveler choice; that the "Other Social Goals" column represents a composite score based on safety, economic efficiency, environment, and equity; and that the "Implementation Obstacles" column represents a composite score based on interest-group concerns, general political obstacles, and institutional or jurisdictional challenges.

The Short List: 13 Strategies for Reducing Congestion and Improving Transportation Options in L.A. County Within Five Years

While the strategy assessments summarized in Figure S.6 provide helpful insights into the advantages and limitations of available options, the fact that they are qualitative—that is, that they are based to some degree on judgment and interpretation—suggests that it would be inappropriate to rely solely on the ratings in developing specific recommendations for Los Angeles. Instead, we first returned to the general insights about congestion and to the specific challenges faced in Los Angeles reviewed earlier in this summary. Based on this information, we constructed an integrated policy framework offering the greatest prospects for reducing congestion and improving transportation options within the L.A. context. With this framework in place, we then relied on individual strategy ratings (as well as the supporting research on which the ratings were based) for guidance in selecting options that would best support the broader aims embedded in the policy framework.

A Policy Framework for Reducing Congestion and Improving Transportation Alternatives in Los Angeles

In reviewing and integrating general insights about the phenomenon of congestion—its causes, its behavior, and its potential cures—as well as the specific contextual challenges faced in Los Angeles, several key themes emerged.

Absent significant policy intervention, congestion in Los Angeles will likely worsen in the coming years. Congestion results from an imbalance in the supply of and demand for road space. For several reasons—including diminished fiscal capacity and resistance among affected communities—significant expansion of the existing road network is unlikely. Notwithstanding the current economic downturn and run-up in fuel prices, meanwhile, forecasted growth in population, the economy, and goods movement suggest that automobile and truck travel will continue to expand in the longer term. Given the nonlinear relationship between traffic volume and travel speed, even small increases in peak-hour driving could make congestion in Los Angeles much worse. The bottom line, then, is that we should be concerned about not just the short-term effectiveness of strategies but the longer-term effectiveness as well.

Managing the demand for peak-hour automotive travel offers the greatest prospects for reducing congestion in Los Angeles. Los Angeles already has a very dense road network that is operated at a high level of efficiency. Significant opportunities to expand or enhance the efficiency of the road network are therefore unlikely. On the other hand, residents of Los Angeles drive a lot relative to the regional population density of the region. Finding ways to manage the demand for driving during the peak hours thus appears to be the most promising—indeed, perhaps the only realistic—way to reduce congestion.

Only pricing strategies can effectively manage automotive demand in the longer term. While many strategies can help reduce congestion in the short term, only pricing strategies resist the effects of triple convergence. And provided that prices are allowed to rise as needed in response to general increases in the demand for automotive travel in future years, pricing strategies will continue to be effective in reducing congestion in the longer term as well. Pricing will also facilitate more-efficient use of existing capacity and raise needed revenue for transportation investments in the region. The latter is especially important given the decline in federal and state excise fuel-tax revenues.

Significant alternative transportation improvements will also be essential. Certain forms of pricing may introduce concerns regarding the ability of lower-income drivers to pay the resulting charges. To mitigate such concerns, it will be essential to offer significantly improved (faster, more reliable, and more convenient) transit options throughout the region. The density of population in Los Angeles is another argument for improving alternatives; with greater density, as argued, it becomes much more difficult to preserve unfettered automobility. Fortunately, it also becomes more realistic to accommodate a greater share of trips via transit, biking, or walking.

Taken collectively, these themes suggested an integrated policy framework offering the greatest prospects for reducing congestion and improving transportation options in Los Angeles. The framework incorporates the following three elements:

- Rely on pricing to manage peak-hour automotive demand, raise needed revenue, and promote more-efficient use of existing capacity.
- Significantly improve transit and other alternative modes.
- Continue to improve the efficiency of the road network but with a shift in emphasis from moving cars to moving people.

Selecting Strategies to Support the Integrated Policy Framework

The next step was to identify strategies that could support the aims of the integrated policy framework most effectively. Here, we relied on the individual strategy ratings for guidance, though we focused on different criteria depending on the role that a given strategy might play within the framework. For example, with pricing strategies, we were especially interested in longer-term performance in reducing congestion, as well as revenue implications, while for alternative transportation and TSM measures, we placed greater emphasis on such factors as cost implications, shorter-term effectiveness in reducing congestion (for TSM options in particular), and other transportation goals, such as mobility, accessibility, and traveler choice. For all strategies, we looked at the current degree of implementation in Los Angeles to understand whether there might be further benefits resulting from additional investment.

We also considered the level of implementation obstacles, such as political opposition, that each of the strategies might be expected to face. Note, however, that a high rating for implementation obstacles would not necessarily rule out the inclusion of a particular strategy; as discussed earlier, most of the easy strategies have already been implemented, and it is expected that many of the more promising strategies are likely to raise at least some political challenges.

A final factor in the analysis was the manner in which different strategies interacted in either complementary or contradictory ways (note that the many possible interactions among 28 different strategies, discussed at length in the appendixes of this book, are difficult to summarize succinctly and thus are not included in Figure S.6). In short, our goal was to select a set of strategies that mutually complemented and reinforced one another.

Based on our analysis, we arrived at a total of 13 recommended strategies for Los Angeles (one of the recommendations spans three distinct strategies, so in total, we selected 15 of the 28 options considered). Of these, 10 appear unambiguously compelling—offering benefits that easily exceed costs—and can be pursued without delay. These are described as our *primary* recommendations. The remaining three also appear quite promising but are subject to some uncertainty: They either require additional study or depend on the outcomes of events currently unfolding in Los Angeles. We thus describe these as our *contingent* recommendations.

Table S.1 introduces, in summary form, the 13 recommendations. The table also indicates the principal goals from the integrated policy framework that each recommendation is intended to support. The table lists the primary recommendations first, followed by the contingent recommendations. Note that we have not attempted to prioritize further among the recommendations within each of the two broad categories; rather, the order of presentation simply follows the same sequence from Figure S.6: TSM measures followed by TDM measures (including pricing) followed by alternative transportation measures.

That a given strategy was not included as a recommendation does not indicate that it lacks merit. There were, in fact, four broad reasons that certain strategies were excluded from the final list:

Table S.1
Strategy Recommendations and Policy Objectives

Recommended Strategy	Manage Peak-Hour Automotive Travel	Raise Transportation Revenue	Improve Alternative Transportation Options	Use Existing Capacity More Efficiently
Primary recommendations				
1. Improve signal control and timing where deficient			○	●
2. Restrict curb parking on busy thoroughfares			○	●
3. Create a network of paired one-way streets			○	●
4. Promote ride-sharing, telecommuting, and flexible work schedules	○		●	
5. Develop a network of HOT lanes	●	○	●	●
6. Implement variable curb-parking rates in commercial centers	●	●		●
7. Enforce parking cash-out law at the municipal level	●			

Table S.1—Continued

Recommended Strategy	Manage Peak-Hour Automotive Travel	Raise Transportation Revenue	Improve Alternative Transportation Options	Use Existing Capacity More Efficiently
8. Promote deep-discount transit passes		○	●	
9. Expand BRT with bus-only lanes			●	
10. Implement a regionally connected bicycle network			●	
Contingent recommendations				
11. Evaluate incident management for the arterial network				●
12. Evaluate potential applications of cordon congestion tolls	●	●		○
13. Levy local fuel taxes to raise transportation revenue	○	●		

NOTE: ○ = an additional goal that a recommendation may help support. ● = a primary goal that a recommendation is intended to accomplish.

- Based on the evidence, the strategy did not appear to be effective in accomplishing its aims.
- The strategy appeared to be effective, but other options for achieving the same goals offered even greater promise.
- The strategy appeared to be effective, but the current level of implementation in Los Angeles is sufficiently advanced that further investment would likely yield diminishing returns.
- The strategy would work at cross-purposes with other options included in the final set of recommendations.

We now summarize the 10 primary and three contingent recommendations developed in this book. For each, we review the key motivating factors for including the strategy, outline likely obstacles that may arise, and discuss possible implementation steps.

Primary Strategy Recommendations

Recommendation 1: *Prioritize and fund investments in upgraded signal timing and control in cities where the current technology is deficient, coordinate signal timing between jurisdictions, and ensure that newly installed technology enables signal prioritization for BRT.*

- ***Motivation:*** Improving the timing and control of traffic signals can lead to significant reductions in delays at arterial intersections, and many cities in the region—most notably the City of Los Angeles— have already invested heavily in this approach. Yet cities with fewer resources often rely on outdated and much less effective timing and control mechanisms, contributing to congestion on the arterial network. Additionally, signal timing and control are often uncoordinated between different cities, leading to bottlenecks at city boundaries. Upgrading signal timing and control technology and improving interjurisdictional coordination will help to reduce delays and will also create the opportunity to implement signal prioritization for faster bus service.
- ***Implementation obstacles:*** The primary obstacle is funding, as the acquisition and installation of signal timing and control technology is very expensive. To illustrate, the City of Los Angeles

recently received $150 million for traffic-signal synchronization improvements from state bond revenues. This will enable the City of Los Angeles Department of Transportation (LADOT) to add 1,117 signals to its current signal-control system, upgrade an additional 1,256 signals to the most recent technology, and add 340 traffic-tracking cameras to help monitor intersections.

- ***Implementation steps:*** Signal timing and control is most deficient in cities that lack sufficient resources to invest in the most up-to-date technology. We therefore recommend that this effort be funded and coordinated at the county level, likely by the L.A. County Metropolitan Transportation Authority (Metro). The initial implementation steps will include (1) conducting a countywide survey of existing signal technology and current traffic conditions on the arterial network so as to prioritize investments and (2) identifying or developing a revenue stream to fund these investments. Follow-on steps will include the provision of priority-based funding and technical assistance to cities for the installation and integration of signal timing and control technology. As a condition for receiving this assistance, we also recommend that cities be required to support efforts to coordinate signals at municipal boundaries and allow for signal prioritization to support more-effective BRT service.

Recommendation 2: *Restrict peak-hour curb parking on all congested thoroughfares and dedicate the additional capacity to bus-only lanes where merited.*

- ***Motivation:*** Prohibiting curb parking on congested thoroughfares during the peak hours can reduce delays by increasing available lane capacity at little cost. We recommend restricting curb parking on all major arterial corridors throughout the region subject to recurring congestion and actively enforcing the restrictions (through prompt ticketing and towing) to ensure their effectiveness. In corridors where the curb lane could carry more total passengers in bus-only mode, we further recommend restricting the lane to bus-only travel during peak hours.

- ***Implementation obstacles:*** Businesses may object that curb-parking restrictions will reduce parking capacity for their patrons, while neighboring communities may be concerned that more vehicles will need to park on the surrounding residential streets. Applying this strategy systematically on all large boulevards throughout the region will reduce perceptions that certain businesses or neighborhoods are being treated unfairly. Another option for overcoming opposition may be to install parking meters in surrounding residential neighborhoods for the use of visitors under the condition that resulting revenues would be allocated to public improvements that would benefit local residents (e.g., repairing sidewalks, burying utility lines, or creating pocket parks). This could create enthusiasm among surrounding neighborhoods for the idea of allowing visitors to pay for the privilege of parking on their streets, yielding a win-win result.
- ***Implementation steps:*** The first step will be to identify congested major arterial thoroughfares that allow curb parking during peak hours. At the same time, corridors with significant bus service should be evaluated for the potential to create bus-only lanes. The next step will be to implement curb-parking restrictions systematically, creating bus-only lanes where merited.

Recommendation 3: *Develop a network of paired one-way streets in high-volume travel corridors throughout the region.*

- ***Motivation:*** Converting streets from two-way to one-way operation will increase road capacity, as the median strip will no longer be required (i.e., it can be converted to a travel lane). It will also reduce delays by enabling progressive signal timing in the direction of travel and making it easier and faster to make left turns. We also recommend the development of bus-only lanes and bicycle lanes on the system of one-way streets.
- ***Implementation obstacles:*** Local businesses may prefer two-way streets with slower-moving traffic so that potential customers will not speed past a store without seeing it. One-way-street proposals are often combined with curb-parking restrictions, and

local businesses may oppose this as well. Owners of surrounding homes may fear more cut-through traffic when vehicles need to backtrack, and they may also be concerned that higher travel speeds will diminish the pedestrian environment and introduce safety risks. While our review of the evidence suggests that one-way streets tend to be safer than two-way streets, evidence on increased neighborhood cut-through traffic is less definitive.

- ***Implementation steps:*** The first recommended step is to implement paired one-way streets on Pico Boulevard and Olympic Boulevard, a proposal that has already received considerable attention, with the provision that a lane be dedicated to bus-only service in the peak hours (see the plan by Rifkin, 2007). While implementing one-way streets with bus-only lanes on Pico and Olympic, city planners can examine other corridors where paired one-way streets might offer similar opportunities. Given the density of arterial traffic on the Westside (see Figure S.4), this should be an area of initial focus. Assuming that the Pico/Olympic project proves successful, the implementation of other paired one-way streets would then proceed.

Recommendation 4: *Bolster outreach efforts to promote voluntary TDM—including ride-sharing, telecommuting, and flexible work schedules—at businesses and other large organizations throughout the county.*

- ***Motivation:*** Within the context of our integrated policy framework for Los Angeles, the implementation of pricing strategies will require improved alternatives to driving alone during peak hours. The three strategies encompassed by this recommendation—ride-sharing, telecommuting, and flexible work schedules (e.g., four 10-hour days per week)—all present viable options for either avoiding peak-hour driving or not driving at all. Current media reports suggest that with the recent run-up in fuel prices—which should have an effect on driving behavior similar to that of pricing strategies—many auto commuters are already seeking other options to save on their gasoline bills. Our recommendation is that a regional agency, such as Metro, devote greater funding to assist employers in developing voluntary TDM programs to create additional commuting options for their employees. We also recommend expanded support for related programs, such as dynamic ride-matching.

- ***Implementation obstacles:*** Our purpose in recommending enhanced promotion of ride-sharing, telecommuting, and flexible work schedules is not to manage the demand for peak-hour automotive travel (though it may help support that aim); that role is filled instead by our pricing recommendations. Rather, within the context of our integrated policy framework, these traditional TDM measures are intended to provide travel options that would enable commuters to avoid paying higher fuel prices as well as the tolls that would be applied with some of the pricing strategies. Accordingly, our recommendation calls for voluntary rather than mandatory measures. Political obstacles are thus not likely; rather, the primary obstacle will be related to securing the necessary funding for the effort.
- ***Implementation steps:*** One of the first steps will be for Metro (or any other agency that chooses to take the lead on this) to review available research and best practices to develop a toolbox of approaches that employers can implement to increase ride-sharing, telecommuting, and flexible work schedules for their employees. The potential to enhance dynamic (ad hoc) ride-sharing options—among commuters as well as other travelers—also merits further attention. A second initial step will be to secure additional funding to support outreach efforts to employers as well as to further subsidize specific TDM programs already under way (such as the CommuteSmart regional ride-matching program and Metro's existing vanpool program). Follow-on activities will then consist of direct outreach and interaction with employers to provide assistance for ride-sharing, telecommuting, and flexible-schedule options for their employees.

Recommendation 5: *Develop a network of HOT lanes on freeways throughout the county and apply any net revenue to the subsidization of express bus service in the HOT lanes.*

- ***Motivation:*** HOT lanes carry both carpools and single-occupant vehicles (SOVs). Solo drivers pay a toll that varies with demand to ensure that the lanes always remain free-flowing, while carpools, depending on the policies adopted, may travel for free, pay a reduced toll, or pay the full toll (even in the latter case, however, carpoolers still save on tolls on a per-person basis). HOT lanes offer several key benefits. They provide a valuable opportunity to pay for faster and more reliable travel time for trips (such as to the airport, to a doctor's appointment, or to day care) when there is a high value for arriving on time. Evidence from existing HOT lanes shows that drivers from all income groups value and make use of this option. By maintaining free-flowing travel speeds, HOT lanes also carry far more vehicles per hour than congested free lanes do, thus enhancing the capacity of the freeway without adding more lanes. As an additional benefit, buses can use HOT lanes, thus improving transit services in the region. The fact that congestion is severe on most highways in the county (see Figure S.3) serves as motivation for developing a network of HOT lanes rather than one or two stand-alone applications.
- ***Implementation obstacles:*** Because HOT lanes offer many benefits, and because use of the HOT lanes is optional rather than mandatory, this strategy enjoys a high degree of public support in places where it has been implemented (Sullivan, 2000; Supernak et al., 2002). Within Los Angeles, however, two implementation obstacles will need to be overcome. First, some freeways do not have high-occupancy-vehicle (HOV) lanes that could be converted to HOT lanes. Second, many of the existing HOV lanes are already operating at or near full capacity during the peak hours such that there would be little space to sell to SOVs. To develop a full network of HOT lanes, it will therefore be necessary to consider such options as (1) converting current general-purpose lanes to HOT lanes, (2) increasing the number of passengers required

(e.g., from two or more to three or more) to qualify as an HOV in order to free up additional space in the lane, or (3) requiring that both SOVs and carpools pay the toll (note that the desire to split a toll among multiple vehicle occupants may continue to encourage the formation of carpools). Each of these actions is likely to face opposition. Enabling state legislation will also be required.
- ***Implementation steps:*** We recommend beginning with a HOT-lane demonstration project on the Interstate 110 (I-110) Harbor Transitway facility to familiarize L.A. residents with the benefits of this strategy. The Harbor Transitway, an 11-mile HOV and bus facility opened in 1996, is an ideal initial candidate, as it already includes two lanes flowing in each direction, which allows for smoother operation. The next step will be to develop a plan to provide HOT lanes on all congested freeways in the county. This may require, as noted, the conversion of existing general-purpose lanes in some cases and the modification of HOV passenger limits in others.

Recommendation 6: *Implement variable curb-parking charges in all busy commercial and retail districts, returning a share of the revenue to local merchants to invest in public amenities and using the remainder to fund municipal transportation investments.*

- ***Motivation:*** Varying parking charges by location and time of day to ensure that there will usually be one or two free spaces available on any block will lead to significant reductions in congestion in busy commercial and retail areas by eliminating the need for visitors to drive around searching for an available space. If implemented broadly throughout the region, net revenues will likely fall in the range of tens to hundreds of millions of dollars annually.
- ***Implementation obstacles:*** Implementing variable curb-parking rates may require investment in new meter technology, but increased parking revenues should easily offset the initial capital cost. Retailers may also fear that higher parking charges will drive customers away; though this proves not to be the case (Shoup, 2005), returning a share of the increased parking revenue to local

merchants for investments in public amenities can still be useful in overcoming this potential opposition.

- ***Implementation steps:*** Cities can pursue this action immediately. Key steps include installing the necessary metering technology and developing a city ordinance specifying that prices will be allowed to vary such that a few spaces remain vacant on each block. The legislation should also establish that local districts will receive a portion of the resulting revenue for investing in public improvements.

Recommendation 7: *Enforce the existing California parking cash-out law at the municipal level in cities where a significant share of employers lease parking.*

- ***Motivation:*** With parking cash-out, workers whose employers lease parking on their behalf are given the option of receiving cash in lieu of free parking. This creates a strong financial incentive to carpool, walk, bike, or take transit instead of driving alone to work, thereby reducing the number of commuters on the road.
- ***Implementation obstacles:*** California requires that firms with more than 50 employees who lease parking for their employees offer the cash-out option, but the law is not enforced. We are recommending that cities take the necessary steps to enforce the law in place of the state. The primary challenges are likely to fall in the legal and administrative categories, but it is not anticipated that they will be difficult to overcome. Importantly, businesses are unlikely to oppose this strategy because (1) the parking cash-out option is viewed as a valuable employee benefit and (2) it costs little to implement, since the law applies only to firms that lease parking and can choose to offer the cash instead of paying the lease (Shoup, 1997).
- ***Implementation steps:*** Cities can pursue this strategy immediately. The City of Santa Monica already requires parking cash-out for qualifying employers, and the City of Los Angeles is exploring a promising approach under which employers that lease parking would be required to offer parking cash-out as part of

the business-permitting process. Parking cash-out should lead to significant reductions in the number of employees who drive alone to work, and we further recommend that parking cash-out programs be carefully monitored to quantify these effects. The resulting information can then serve as a basis for reducing off-street parking requirements for real-estate developments in which parking will be leased. Over time, as more office buildings with leased parking are developed, the number of employers able to offer parking cash-out will increase, thus expanding the benefits of the program.

Recommendation 8: *Develop and aggressively market deep-discount transit fares to employers in areas that transit serves well.*

- ***Motivation:*** Deep-discount-fare programs enable large organizations (such as universities or firms) to purchase transit passes for all members or employees at significant discounts. When structured properly, they lead to increased transit ridership, increased transit-operator revenues, and reduced transit operating deficits. And from the purchasing organization's perspective, they are often cheaper than providing additional parking or other transportation benefits. Whereas pricing strategies recommended in this book make driving more expensive, deep-discount passes reduce the cost of transit, thereby making it even more attractive by comparison. Successful deep-discount programs already exist in the L.A. region but have been applied only to a limited extent.
- ***Implementation obstacles:*** The principal obstacles are administrative and should not be difficult to overcome.
- ***Initial implementation steps:*** Metro and other transit providers in the county can begin efforts to develop and more aggressively market deep-discount-fare programs immediately.

Recommendation 9: *Expand BRT in urban areas with dedicated bus-only lanes on the arterial network and express freeway service in HOT lanes.*

- ***Motivation:*** This recommendation will significantly improve the speed, convenience, and reliability of public transit in Los Angeles at relatively low cost (in comparison with current estimates of $400 million or more per mile to construct subways). The improved transit options will provide viable alternatives for those wishing to avoid the higher price of automotive travel that would result from pricing strategies and will benefit the many L.A. residents who already rely on the bus system for their daily travel needs. Thus, current transit users, new transit users, and those driving on less congested roads will all be better off. While the Metro Orange Line busway in the San Fernando Valley includes 14 miles of exclusive right-of-way, Metro Rapid bus lines share the streets with general traffic and must therefore suffer the same congested travel conditions. Metro Rapid lines include many features of the BRT concept that has taken the transit world by storm in recent years, including signal prioritization, more-frequent service, limited stops, real-time next-bus information at stops, and so on. These features have improved BRT travel speeds by about 20 to 30 percent over conventional local bus service (Metro, 2000). Yet absent partially or fully reserved rights-of-way, service is still slow in many places; the average daytime travel speed for Rapid buses along the Wilshire corridor, for example, is just 11 mph (Jeff, 2007a).
- ***Implementation obstacles:*** Allowing buses to travel in HOT lanes on the freeway network is a promising opportunity. On the arterial system, however, the primary obstacle to providing bus-only lanes is that many drivers are likely to argue that *all* vehicles should be allowed to use the lanes. Provided that a bus-only lane will facilitate greater total passenger (as opposed to vehicle) throughput in a corridor, however, reserving a lane for the exclusive use of bus transit represents the most efficient use of the capacity.
- ***Implementation steps:*** We recommend implementing bus-only lanes on Wilshire Boulevard in the curb lane during peak travel hours—an idea already being evaluated by the City of Los Angeles and Metro—as a first step in developing a regional network of dedicated bus-only lanes on the arterial system. Bus service

along the Wilshire corridor already accommodates about 100,000 boardings each weekday; with faster travel speeds in a bus-only lane, ridership in the corridor should expand even further. The next steps will be to plan and implement a network of bus-only lanes on the arterial system and add express freeway bus service as HOT lanes are implemented. A key element of the planning effort will be to monitor the effects of the initial bus-only lanes on Wilshire, examining the outcomes with respect to bus travel speed and ridership. Based on this information, it should then be possible to examine ridership on other major arterial corridors and project how it might increase with the introduction of bus-only lanes during peak hours. In cases in which current ridership plus anticipated gains suggest that bus-only lanes will support greater total passenger throughput, the curb lane should be restricted to bus service during peak hours. Additional opportunities to create bus-only lanes may arise with the conversion of two-way streets to one-way operation and the restriction of curb parking on busy thoroughfares during peak hours.

Recommendation 10: *Develop an integrated, regionwide bicycle network, with a specific focus on dense urban areas where bicycles can serve a large share of trips.*

- ***Motivation:*** This recommendation falls under the category of improving alternatives to driving in Los Angeles. While cycling may not serve as a potential replacement for a long commute, it can certainly facilitate shorter trips (and cycling in combination with transit can serve longer trips). Los Angeles offers an ideal climate for cycling throughout much of the year, and many excellent bicycle facilities already exist. Yet the system is not clearly legible (i.e., easy to understand) to existing or potential riders, bicycle routes are not well connected at the regional level, and existing lanes often end abruptly at municipal boundaries. To make cycling a more viable mode of transportation in the region, especially in urban areas, we recommend the development of an integrated and legible regional bicycle network, including additional, clearly identifiable bike lanes and bike paths along with improved bicycling amenities—such as bicycle-storage lockers—at transit hubs. Because bicycling can be more dangerous than driving per mile of travel (Pucher and Dijkstra, 2000), we also recommend additional expenditures on bicycle-safety training programs.

- ***Implementation obstacles:*** The principal obstacle to expanding the bicycle network in Los Angeles is insufficient funding combined with the relatively low prioritization given to bicycling facilities. There may be some institutional obstacles as well; bicycling improvements are typically implemented by individual cities, and not all cities in the county have been willing to devote road space to biking. To overcome concerns that bike lanes would reduce available space for automobiles on busy thoroughfares, it may be useful to consider opportunities for creating bike lanes on less heavily traveled routes.
- ***Initial implementation steps:*** We recommend allocating additional revenue to bicycling facilities and completing Metro's existing Bicycle Transportation Strategic Plan (Metro, 2006) as a starting point. Further opportunities to expand the bicycle network may arise with the conversion of two-way streets to one-way operation.

Contingent Strategy Recommendations

Recommendation 11: *Evaluate the costs and benefits of implementing a regional incident-management system on the arterial network.*

- ***Motivation:*** Traffic incidents—including crashes and vehicle breakdowns—are responsible for a very large share of urban congestion. By facilitating faster detection of, response to, and clearing of such occurrences, incident-management systems can have a dramatic effect on reducing the resulting delays. Metro and the California Department of Transportation (Caltrans) already fund and manage an extensive Freeway Service Patrol (FSP) system that helps to accomplish these goals. We recommend testing a similar program on the arterial network and, should it prove successful, expanding it regionally. *Note that most incident-management* *systems implemented to date have focused on freeway operations, and thus there are few data to indicate how effective such a program would be on the arterial system; for this reason, we treat it as a contingent recommendation, one requiring further evaluation.* (LADOT has reportedly experimented with an arterial-based incident-management program, but, as of this writing, the results of the evaluation were not available.)

- ***Implementation obstacles:*** Should arterial incident management prove to be a cost-effective strategy, the primary obstacle will be to secure funding for the program.
- ***Initial implementation steps:*** The first step, as noted, will be to evaluate the costs and benefits of an arterial incident-management system through some type of trial or demonstration project. If, based on the preliminary evaluation, arterial incident management proves to be cost-effective, the next step will be to implement an ongoing program. While individual cities could pursue this, there may be economies of scale to be achieved through regional implementation. In this case, Metro would likely be the lead agency, as it already manages the county's FSP program.

Recommendation 12: *Evaluate the potential for implementing cordon congestion tolls around major activity centers in the region.*

- ***Motivation:*** Cordon congestion tolls apply a charge for any vehicle that enters or travels within a designated cordon area (for example, a central business district) during peak hours. This leads to significant congestion reductions within the charging zone and confers additional benefits, such as fewer traffic accidents and reduced vehicle emissions. Cordon tolls can also raise hundreds of millions of dollars per year. *To date, however, cordon tolls have been implemented only in relatively large, centralized cities with extremely well-developed transit systems. It remains unclear whether cordon tolls could be equally effective in the L.A. context without creating legitimate equity concerns. We thus categorize cordon tolls as a contingent recommendation and suggest that further research be conducted prior to pursuing implementation.*

- ***Implementation obstacles:*** To implement cordon tolls in Los Angeles, three obstacles will need to be overcome. First, Los Angeles is much more polycentric than other cities where this approach has been employed; as a result, it may be appropriate to consider multiple cordon-toll areas. Second, because cordon tolls (unlike HOT lanes) are required of most drivers who enter the zone, they raise concerns that lower-income drivers might be disproportionately burdened by the cordon charges (Santos and Shaffer, 2004). Such concerns make it necessary to provide high-quality transit service to any area subject to a cordon toll to ensure that there are viable alternatives to driving. While revenues raised by a cordon toll can help fund improved transit services in the longer term, many of the investments will need to occur before the cordon toll is implemented in order to effectively mitigate concerns. Third, the concept of cordon tolls does not enjoy a high degree of public support in Los Angeles, both because it is an unfamiliar idea and because most drivers expect that they would incur significant costs without appreciable travel-time savings. Educational outreach about the effectiveness of this approach along with efforts to build popular and political support would thus be necessary before such a program could be pursued. Choices about how to allocate the resulting revenue will be instrumental in helping to build support.
- ***Implementation steps:*** Cordon congestion tolls are complex, and the first required step will be to study the potential for their application in Los Angeles (note that Metro is currently funding such a study). Interrelated issues to address in the study include where to locate the cordon boundaries, what electronic tolling technology to employ, how to structure the tolls (e.g., flat rate versus varying by time of day), how to allocate the revenues, and how the cordon toll will affect different spatial and demographic segments of society. Given the high levels of congestion on both the freeways and the arterial network between the Westside and downtown Los Angeles, we suspect that such areas as downtown Los Angeles, Century City, Santa Monica, and Los Angeles International Airport (LAX) might emerge as promising candidates for cordon

tolls. Should the study find that cordon tolls would be feasible and beneficial, the next steps will be to conduct outreach, develop the necessary political consensus, and implement the recommended tolling zones. To mitigate equity concerns, fast and efficient transit options serving the charging zones will have to be established *before* the cordon-toll operations commence.

Recommendation 13: *Implement local fuel-tax levies at the county level to raise transportation revenues and (to a lesser extent) reduce the demand for driving.*

- ***Motivation:*** Failure to raise federal and state fuel taxes in recent years to keep pace with inflation and improved fuel economy has resulted in worsening transportation-funding shortfalls in Los Angeles and elsewhere. Los Angeles has, in the past, relied on sales-tax ballot measures to raise local transportation funds, but fuel taxes offer more compelling benefits. With fuel taxes, the amount one pays is based (roughly) on the amount one drives, and aligning costs and benefits can be viewed as more equitable (Wachs, 2003b). Fuel taxes also act as incentive to drive less, and this can help to reduce congestion. Further, fuel taxes encourage the purchase of more fuel-efficient vehicles, thus leading to important environmental benefits. Neither sales taxes nor other general-revenue sources offer these same advantages. To promote a more stable revenue stream and prevent the erosion of revenues against future inflation, we would recommend that local fuel taxes be structured on a cents-per-gallon basis and indexed to increase with either the consumer price index or the construction cost index. *Note that Metro is already taking steps to place an additional 0.5-percent sales tax to fund transportation improvements on the ballot in the fall of 2008, and it is unlikely that the county would simultaneously pursue a sales-tax measure as well as a local fuel-tax levy. Should the sales-tax effort stall, we would then recommend the development of a local fuel-tax measure in its stead. This leads us to categorize this as a contingent recommendation.*

- ***Implementation obstacles:*** Enabling state legislation would be required before local fuel taxes could be levied in Los Angeles. Assembly Bill (AB) 2558, recently introduced by Assemblymember Mike Feuer, could serve this purpose. AB 2558 would allow L.A. County to levy a carbon tax of up 3 percent of the fuel purchase price (or roughly $0.12 to $0.15 per gallon at current fuel prices); though this represents a slight departure from our recommendation of structuring the tax on a cents-per-gallon basis indexed to rise with inflation, the broad intent of the measure is otherwise similar. Voters would then be required to approve the tax. Note, however, that we are using the term *tax* in the generic sense. From a technical perspective, Assemblymember Feuer has, in fact, attempted to structure AB 2558 as a user fee rather than as a tax (Newton, 2008). A key difference is that, under California law, user fees require only 50-percent voter approval (although there are more restrictions on the allocation of the revenue), whereas taxes require a two-thirds vote. Yet despite recent evidence that voters express growing support for efforts to link transportation funding with environmental goals (Dill and Weinstein, 2007), gaining even 50-percent voter approval may prove to be politically challenging given that (1) the economy appears to be weakening, (2) the cost of gasoline has been rising dramatically, (3) additional taxes will be viewed as burdensome for lower-income drivers, and (4) Metro is already pursuing a sales-tax measure for the current ballot cycle.
- ***Initial implementation steps:*** The first step, as noted, will be to pass enabling state legislation. The second step will be to gain voter approval, and this will require concerted efforts to build political support.

The 13 Recommendations, Collectively, Should Produce Substantial Benefits for Los Angeles

As discussed earlier, we did not examine available strategies in terms of project-specific details, nor did we rely on modeling for the analysis.

This precluded the ability to make specific predictions on the degree to which travel speeds might be improved and congestion reduced through the implementation of the 13 recommendations. Even so, there is ample evidence on how the strategies perform in cities where they have already been applied. If we assume that the strategies will work as well in Los Angeles as they have elsewhere (and we have no reason to suspect otherwise), then the benefits afforded by the recommendations should be considerable indeed. Here, we summarize evidence on the likely effectiveness of the recommendations in terms of reducing congestion, improving transportation alternatives, and raising needed revenue.

Increasing Travel Speed and Reducing Delays on the Road Network

- Signal timing and control investments in areas where the current technology is lacking or deficient can increase travel speeds by 10 to 22 percent and reduce travel times by 8 to 17 percent on the arterial network (LADOT, undated[b]; MITRE Corporation, FHWA, and USDOT, 1996).
- Peak-hour curb-parking restrictions with active enforcement can improve arterial travel speed by around 10 percent (Kumar, 2007).
- Paired one-way street conversions can increase travel speed by about 20 percent and reduce travel time by 20 to 30 percent (Cunneen and O'Toole, 2005; Rifkin, 2007; Stemley, 1998; TTI, 2001).
- HOT lanes can maintain free-flowing travel speeds (60 to 65 mph) during peak travel hours while carrying up to twice the volume that congested general-purpose lanes do (Obenberger, 2004).
- Cordon tolls can reduce traffic volume within the charging zone by about 15 percent, increasing average travel speeds by up to 100 percent (Fabian, 2003; Goh, 2002; Santos and Shaffer, 2004).
- Variable curb-parking rates can reduce traffic volumes in busy commercial and retail districts by about 30 percent on average and by up to 90 percent in at least one example (Shoup, 1997).

Making Alternative Transportation More Attractive and Convenient

- Parking cash-out can provide sufficient financial incentive for about 15 percent of employees to shift from driving to alternative modes (Shoup, 1997).
- Deep-discount transit fares can lead to transit ridership gains of up to 200 percent among members of participating organizations (Brown, Hess, and Shoup, 2001).
- BRT featuring bus-only lanes can result in much faster transit service at relatively low cost (Levinson, Zimmerman, Clinger, Rutherford, Smith, et al., 2003); evidence from the Metro Rapid system demonstrates that reductions in travel time stimulate proportional gains in ridership (Metro, 2000).

Raising Needed Transportation Revenue

- HOT lanes can raise sufficient revenue to subsidize express bus operations within the corridor (Obenberger, 2004; Poole and Orski, 2000; SANDAG, 2007).
- Cordon tolls can generate tens to hundreds of millions of dollars in annual net revenue (ECMT, 2006).
- Variable curb-parking rates can raise well over $1 million in annual net revenue within a single urban retail district (Shoup, 2005). Implemented throughout the region, the strategy would likely raise tens or even hundreds of millions of dollars for municipalities in the region.
- A local fuel tax on the order of $0.12 to $0.15 per gallon in L.A. County would generate around $500 million in annual revenue ("Feuer Introduces Package of Transportation Bills," 2008).

These benefits are likely to accrue throughout many areas of L.A. County. While some of the strategy recommendations are tailored especially to dense urban areas, others promise to be equally effective in urban or suburban settings. Table S.2 provides a summary of the types of facilities (arterials versus freeways) and areas (suburban, urban,

Table S.2
Scope of Effects

Recommended Strategies	Facility		Area		
	Arterial	Freeway	Suburban	Urban	Commercial Center
Primary recommendations					
Signal timing and control	•		•	•	•
Curb-parking restrictions	•			•	•
One-way streets	•		•	•	•
Voluntary TDM	•	•	•	•	•
HOT lanes		•	•	•	•
Variable curb-parking rates	•				•
Parking cash-out	•	•		•	•
Deep-discount transit passes	•	•		•	•
BRT with bus-only lanes	•	•		•	•
Regional bicycle network	•			•	
Contingent recommendations					
Arterial incident management	•		•	•	•
Cordon congestion tolls	•	•			•
Local fuel taxes	•	•	•	•	•

or commercial center) in which we would expect the effects of each of the strategy recommendations to be most pronounced.

Key Dependencies and Interactions

We recognize that, as leaders consider the recommendations in this book, there may be a desire to focus only on those that seem to offer significant benefits without stirring political controversy. Against this possibility, we would like to reiterate two crucial observations to bear in mind. First, as indicated in the summary of benefits, both TSM and pricing strategies will help to reduce delays and improve travel speeds. Given the shorter-term effects of triple convergence, however, as well as the longer-term effects of general increases in the demand for automotive travel, the congestion-reduction benefits of TSM strategies are, in the end, likely to be short-lived. If Los Angeles hopes to achieve traffic-congestion reductions that are sustainable for the longer term, it is essential that the set of adopted policies include pricing strategies.

Second, though many of the strategies we have recommended would be highly effective if implemented on their own, there are many ways in which the recommendations, taken together, either complement or depend on one another. Stated another way, the whole of the recommendations can be considered as greater than the sum of its parts. While it may not prove possible to implement all of the strategies recommended here, Los Angeles would benefit greatly by implementing as many as possible as part of a coordinated and comprehensive effort. In the following paragraphs, we highlight three issues in which the complementarities and interdependencies among the 13 primary and secondary recommendations appear to be especially critical.

Funding: Many of the recommended strategies, but especially signal timing and control, extensive one-way-street conversions, arterial incident management, and BRT expansion, will require significant additional revenue. Cordon congestion tolls, variable curb-parking rates, and local fuel taxes all offer the potential to raise significant county or municipal transportation revenues to help in this regard. HOT lanes and deep-discount transit-fare programs can also provide modest net revenue.

Ability to pay: Cordon tolls—and, to a lesser extent, HOT lanes, variable curb-parking rates, and local fuel taxes—are likely to raise fairness concerns about the ability of lower-income drivers to afford the resulting charges. To mitigate these concerns, it will be essential to improve nonautomotive travel alternatives in the region through such strategies as voluntary employer trip-reduction programs, deep-discount transit fares, and enhanced BRT service featuring bus-only lanes on the arterial network and express bus service in HOT lanes on the freeways.

Competition for road space: One of the most promising short-term strategies for improving the speed and convenience of transit in Los Angeles is the creation of bus-only lanes in transit-rich arterial corridors, such as Wilshire Boulevard. Yet if these corridors are already congested, drivers are likely to object strenuously to the allocation of an existing lane to bus-only service, even if such treatment would facilitate greater overall passenger (as opposed to vehicle) throughput. Peak-hour curbside-parking restrictions and one-way-street conversions are both valuable in this context because they create *additional* lane capacity—that is, capacity not already claimed for general-purpose traffic. As a result, it may be easier, politically, to create bus-only lanes in tandem with curbside-parking restrictions or one-way street alignments than it would be to create bus-only lanes without these other changes.

Next Steps for L.A. County

While individually and collectively promising, each of the strategy recommendations involves change from the status quo and thus is likely to engender resistance from some group of stakeholders. So how to begin? What must Los Angeles do to implement these recommendations and get on with relieving the region's congestion problems?

Table S.3 examines this question from a jurisdictional perspective, summarizing the roles that various governmental entities would likely need to fulfill in implementing the recommendations. Note that many of the strategies could involve multiple actors. *Lead* indicates that a particular governmental entity would likely assume leadership for

Table S.3
Public-Sector Implementation Roles

Recommended Strategies	State Legislature	Caltrans	Metro as RTPA	LACDPW	Cities (DOTs)	Transit Operators
Primary recommendations						
Signal timing and control			Lead	Required	Required	
Curb-parking restrictions					Lead	
One-way streets			Optional		Lead	
Voluntary TDM			Lead		Optional	
HOT lanes	Required	Required	Lead		Optional	
Variable curb-parking rates					Lead	
Parking cash-out					Lead	
Deep-discount transit passes						Lead
BRT with bus-only lanes					Required	Lead
Regional bicycle network			Lead		Required	
Secondary recommendations						
Arterial incident management			Lead		Optional	
Cordon congestion tolls	Required	Required	Required		Lead	Required
Local fuel taxes	Required		Lead			

NOTE: RTPA = regional transportation planning authority. The "Metro as RTPA" column refers to the agency's broad planning function—that is, Metro as the RTPA for L.A. County. Metro's additional role as the region's largest transit provider is captured in the "Transit Operators" column. LACDPW = L.A. County Department of Public Works. DOT = department of transportation.

implementing a recommendation, while *Required* means that the entity would need to support implementation efforts. Finally, *Optional* indicates that an agency's involvement might be helpful but not essential.

Although a large share of the recommendations will require coordination among multiple agencies or jurisdictions, developing the necessary political support among residents and elected officials will likely prove an even greater challenge. For this reason, we also examined complementary strategies for helping to build political consensus to support transportation-policy reforms in Los Angeles. We began by reviewing the scholarly literatures in planning and political science for relevant insights into the ingredients of successful collective action. We next spoke with local elected officials, agency staff, and leading transportation scholars and practitioners for further guidance. By integrating their advice with the general guidelines from the literature, we arrived at 10 complementary recommendations for helping to build the political consensus needed for many of the congestion-reduction strategies described in this book.

Note that one of the key ingredients in building political support is strong and persistent leadership, a role that can, in principle, be filled by either elected officials or community groups. We argue that the latter option offers the greater prospects for Los Angeles, as community leaders do not face the same constraints—such as election cycles and term limits—as elected officials do. With that in mind, the first six general recommendations would be relevant to community leaders who would like to assume a role in building support for the strategies recommended in this book. The final four recommendations focus on programmatic design considerations for specific congestion-reduction or revenue strategies and may be most helpful to elected officials and agency staff members in their efforts to plan and implement policies that can attract and maintain public support.

General Recommendations for Building Political Support

Form a coalition of community representatives to provide political leadership. The coalition would help to build political support and provide ongoing encouragement for elected leaders across L.A. County to pursue a coordinated and aggressive transportation agenda.

Include diverse interest groups when forming the community coalition. For the leadership of the community coalition to be viewed as legitimate, it must include the full range of affected parties. Examples include key industry groups; racial and ethnic groups; automotive, transit, and bicycle advocates; and environmental and social-justice advocates.

Develop agreement on the need for aggressive action to halt growth in congestion. To gain broad support for potentially controversial recommendations, it will be necessary to build agreement that L.A. traffic congestion has reached unacceptable levels, that it will likely worsen in future years, that existing remedies are ineffective, and that new strategies, including the use of pricing, are needed.

Define, broadly, the problems associated with congestion to help foster agreement on the need for action. Not everybody will agree that congestion by itself is such a severe problem that something must be done. Though no one enjoys sitting in traffic, some residents might feel that paying congestion tolls and paying higher prices for parking are even less desirable. Highlighting the full range of problems exacerbated by congestion—including economic, environmental, and social concerns—will help build greater support for the argument that aggressive action is needed.

Develop a compelling narrative of the benefits of action. Outlining the consequences of failing to act may be sufficient to motivate some, but offering a compelling vision of the benefits that will result from the comprehensive set of recommendations may persuade even more. Myers (2007) noted that effective leaders are also effective communicators who can weave together both majority and minority goals and preferences to create a community narrative that compels collective action. Myers further argued that an artful mix of future-oriented graphics, metaphors, and storytelling can bolster the success of this narrative.

Develop support for comprehensive programs rather than individual projects. The congestion-reduction strategy recommendations were selected to complement or reinforce one another. For instance, cordon tolls will reduce traffic congestion in crowded urban zones, and they will raise significant revenue for transportation investments. Transit improvements will provide viable options for those wishing to avoid the higher charges for driving during peak hours, and they will offer benefits to those who already rely on nonautomotive modes of travel. One-way street alignments will lead to improved travel speed and reduced intersection delays for automobiles, and they will introduce opportunities for creating bike lanes and bus-only lanes. As a package, the integrated strategy recommendations offer greater combined benefits, spread the costs more evenly, and offset many of the concerns that might apply to specific strategies in isolation.

Recommendations for the Design of Specific Strategies

Apply congestion-reduction strategies systematically. Certain strategies, such as one-way-street conversions and the prohibition of curb parking, may benefit motorists and transit users who travel in a particular traffic corridor but impose costs on local residents and merchants. To reduce concerns that certain groups are being asked to sacrifice for the benefit of others, we recommend implementing strategies systematically throughout the region (following, in some cases, initial demonstration projects to help familiarize residents with the benefits offered by some of the more innovative strategies). This will diffuse the costs and spread the benefits more broadly such that specific neighborhoods and retail districts will not feel that they have been singled out for unfair treatment.

Allocate pricing revenues to enlist support and mitigate equity concerns. Lack of organized support, on one hand, or spirited opposition, on the other, can prevent the implementation of pricing policies. Yet such strategies create a significant revenue stream, and choices about the allocation of the revenues can help to both build support and defuse opposition. For example, three possible allocations of the revenue resulting from cordon congestion tolls might include paying for public amenities within the charging zone, improving transit services in and around the charging zone, and partially subsidizing toll rates for lower-income motorists with demonstrated need (the tolls should not be entirely eliminated for lower-income drivers, as this would defeat the intent of using prices to manage peak-hour travel demand). Such allocations may help to enroll the support (or at least reduce the opposition) of local retailers and social-justice advocates, two of the groups most likely to oppose cordon tolls. Returning a portion of revenues to local merchants for public improvements has also been demonstrated as an effective strategy to build support for variable curb-parking rates (Shoup, 2005). Suitable protections should be instituted to ensure that that funds are expended only on specified purposes.

Allocate gas-tax revenues to build broad geographic support for the measure and ensure that revenues will be dedicated strictly to projects for improving mobility and reducing congestion. Local fuel taxes would likely be implemented at the county level. To build widespread support for this revenue measure, it is thus important to ensure that the benefits are broadly and fairly distributed. Echoing the importance of early stakeholder engagement, Hamideh et al. (2006) reported that local transportation-related ballot measures are more likely to be successful when the public has the opportunity to comment on the specific projects that they would most like to see funded. This may include transit improvements in denser urban areas and road improvements in more sparsely populated outlying areas. Strong protections should be provided against the diversion of funds to nontransportation uses, and recipients should be subject to maintenance-of-effort requirements to ensure that new revenues are treated as additions to current transportation expenditures, not substitutes for them.

Provide strong enforcement for strategies that involve pricing or driving restrictions. Recommended strategies that require strong enforcement include cordon congestion tolls, HOT lanes, variable curb-parking rates, curb-parking restrictions, and bus-only lanes. For strategies requiring that motorists either pay more or change their behavior, resentment will build if there is a perception that others are able to cheat the system (Short, 2004). Strong and consistent enforcement prevents this, and the resulting citation fees can help offset the costs of enforcement.

The Integration of Short-Term and Long-Term Strategies

This book, as noted, focuses on strategies that could be implemented and produce results within a few short years. Yet there are also compelling longer-term strategies that Los Angeles might pursue. Land-use reforms related to zoning, density, parking supply, and the mixing of uses may lead to development patterns less reliant on the automobile, for instance, while major infrastructure investments may help to improve transit options or reduce existing bottlenecks on the road network.

The shorter-term recommendations developed in this book, in addition to offering immediate benefits, should also complement longer-term strategies that Los Angeles might pursue. For instance, such strategies as cordon tolls, fuel taxes, and variable parking charges will produce significant and lasting revenue sources that can help to support future infrastructure investments. The implementation of pricing strategies to manage the demand for automotive travel, in turn, may help increase support for more compact development patterns clustering around high-capacity transit corridors and hubs.

Yet many of the strategies can be viewed as transitional as well. For instance, the recommendation to improve BRT includes the suggestion of implementing bus-only lanes on Wilshire Boulevard, one of the region's busiest transit corridors. Metro is currently evaluating the potential to construct a subway running the length of Wilshire, but, as yet, there is insufficient funding to pursue this option. As an interim step, developing bus-only lanes on Wilshire Boulevard would improve the speed and reliability of transit in the corridor at a much lower cost. It would also stimulate additional transit patronage, and this would reduce the level of subsidization ultimately required to build and operate a subway line under Wilshire Boulevard.

Many of the pricing recommendations in this book may also serve a transitional role. With recent technology developments, it is now possible to equip vehicles with on-board computers featuring Global Positioning System (GPS) receivers and digital road networks. As demonstrated in recent trials in Oregon and the Puget Sound region of Washington, this would enable the development of networkwide pricing applications with charges based on such factors as the number of miles driven, the time and location of travel, and the characteristics (such as size, weight, and pollution emissions) of the vehicle driven (for discussion, see Sorensen and Taylor, 2006). If implemented, networkwide pricing applications could replace many of the shorter-term pricing recommendations discussed in this book, including fuel taxes, HOT lanes, and cordon tolls. Yet networkwide pricing represents an even more significant departure from current policies, and it will likely take time (and much debate) before the public is willing to consider this concept. In the interim, such options as HOT lanes and variable parking charges can help to familiarize L.A. residents with the benefits of pricing strategies, making it possible later to pursue more advanced forms of pricing.

Another issue to consider when evaluating policy options from the longer-term perspective—one that has become especially important of late—is the future trajectory of fuel prices. As of this writing, the cost of gas in Los Angeles is well over $4 per gallon, and, in many locations, the price is closer to $5. While many factors affect prices, continuing growth in the demand for oil in rapidly developing countries, such as China and India, suggest that the price may not drop significantly in the coming years and may well rise even further. At the same time, increasing concern about climate change is stimulating federal and state policy debates over how best to reduce greenhouse-gas emissions in the transport sector and elsewhere. The most promising strategies being considered, including carbon taxes and carbon cap-and-trade systems, would almost certainly lead to further increases in the price of gasoline and diesel.

Over the past six to nine months, as fuel prices have continued to climb and the economy has softened, the level of traffic congestion has declined on some routes in Los Angeles. Given that gas prices may remain at current levels or climb even higher in the coming years, it is reasonable to ask whether the recommendations in this book are really needed. We believe that they still offer significant value. To begin with, while congestion has eased, it has not evaporated, and delays on the busiest routes during peak hours are still severe. The economy, though struggling at the moment, will eventually recover, and the demand for automotive travel seems likely to increase over the longer term with projected population growth and economic expansion. And should fuel prices remain high, automotive companies will likely offer greater numbers of fuel-efficient conventional vehicles as well as innovative options, such as plug-in hybrids and all-electric vehicles, to reduce the amount of fuel that drivers must purchase. Thus the prospects for continued increases in congestion in the longer term, in our view, appear considerable.

Even if higher fuel prices prove sufficient to dampen the demand for automotive travel in the longer term, however, the strategies recommended in this book would still serve a useful role. Efforts to improve nonautomotive travel alternatives, for instance, should take on even greater urgency. And while expensive measures for fine-tuning the operational efficiency of the existing road network (e.g., signal timing and control improvements) might receive less attention, options that simultaneously support faster and more reliable transit service—such as curb parking restrictions or one-way street alignments combined with the implementation of bus-only lanes—would also remain important.

Strategies that rely on pricing to manage peak-hour automotive travel present the most challenging question. If higher fuel prices lead to longer-term reductions in automotive travel, would pricing strategies still be helpful, or would they instead represent an additional burden on residents already struggling to pay their fuel bills? We can offer several reasons that it would still be useful to pursue the congestion pricing recommendations discussed in this book. To begin with, they would provide the necessary revenue for improving transportation alternatives in the region, and higher fuel prices could be expected to increase the demand for such options. In addition, as long as there are enough drivers to cause congestion on the busiest routes during peak hours, pricing strategies would still play a useful role in preventing the buildup of severe congestion. Finally, as higher fuel prices become more burdensome, pricing charges would actually become less burdensome. This is because the charges levied under such strategies as HOT lanes, cordon congestion tolls, and variable curb-parking rates are a direct function of demand—that is, they are set high enough to prevent congestion, but not higher. If higher fuel prices stimulate a reduction in automotive travel, the charges would decline correspondingly. And if fuel prices became so severe that congestion dissipated, the charges would be zero.

In summary, the short-term strategies recommended in this book should complement longer-term strategies that Los Angeles might pursue, and they remain useful and appropriate in the face of sharply higher fuel prices.

Consensus Is Possible, with Benefits for All

The success or failure of congestion-reduction efforts in Los Angeles lies less in the realm of engineering and policy analysis and more in the political arena. The recommendations in this book are based on careful research and draw on evidence from successful implementations elsewhere, yet this is not the first publication to feature recommendations for managing peak-hour automotive travel through the mechanism of pricing. Implementing pricing strategies will be far more difficult than recommending them, and the political challenges will be compounded by the complexity of the transportation decisionmaking environment in L.A. County, in which cooperation among multiple agencies is required, and small but vocal opponents have ample democratic and legal options for derailing the process.

That said, there now appears to be a greater willingness among some elected officials in Los Angeles to consider pricing strategies as a fundamental ingredient for reducing congestion and raising revenue to improve transportation options in the region. As of this writing, the Metro Board of Commissioners has voted to accept federal funding under the U.S. Department of Transportation's Urban Partnership Agreement program to develop several HOT lanes in the county, while the City of Los Angeles has begun to install new parking-meter technology that will allow for the implementation of parking charges that vary with demand. While both of these are consistent with the recommendations in this book, they have not yet proceeded beyond the preliminary planning stages, and greater opposition may arise as plans move closer to implementation. This underscores the importance of community leaders stepping in to help develop broader support for such measures.

Encouragingly, our analysis indicates that successful collective action is possible—with strong and persistent leadership, in a process that includes all interested parties from the start and ensures that all concerns receive due attention. Our analysis leads us to believe that the ends will justify the effort, as reducing congestion should help to improve quality of life, enhance economic competitiveness, reduce greenhouse-gas emissions, improve air quality in freeway-adjacent neighborhoods, and improve mobility for drivers and transit patrons alike.

References

American Public Transportation Association, *2007 Public Transportation Fact Book*, 58th ed., Washington, D.C., May 2007. As of February 28, 2008: http://www.apta.com/research/stats/factbook/

APTA—*see* American Public Transportation Association.

Brown, Jeffrey, Daniel Baldwin Hess, and Donald Shoup, "Unlimited Access," *Transportation*, Vol. 28, No. 3, August 2001, pp. 233–267. As of April 23, 2008: http://www.uctc.net/papers/525.pdf

California Department of Industrial Relations, "California Consumer Price Index," Sacramento, Calif., October 16, 2007.

California Department of Transportation, "Fact Sheet: Important Events in Caltrans History," undated. As of May 11, 2008: http://www.dot.ca.gov/hq/paffairs/about/cthist.htm

———, *Highway Construction Cost Index: 1st Quarter Ending March 2008*, Sacramento, Calif., April 28, 2008.

Caltrans—*see* California Department of Transportation.

City of Los Angeles Department of Transportation, "Street Smart Implementation Program," memorandum on the results of retiming and synchronizing traffic signals along priority arterials, provided by City of Los Angeles Department of Transportation staff, undated.

Cunneen, Michael, and Randal O'Toole, *No Two Ways About It: One-Way Streets Are Better Than Two-Way*, Golden, Colo.: Independence Institute, Center for the American Dream, issue paper 2-2005, February 2005. As of February 19, 2008: http://www.i2i.org/articles/2-2005.pdf

Dill, Jennifer, and Asha Weinstein, "How to Pay for Transportation? A Survey of Public Preferences in California," *Transport Policy*, Vol. 14, No. 4, July 2007, pp. 346–356.

Downs, Anthony, *Still Stuck in Traffic: Coping with Peak-Hour Traffic Congestion*, Washington, D.C.: Brookings Institution Press, 2004.

ECMT—*see* European Conference of Ministers of Transport.

European Conference of Ministers of Transport, "Summary and Conclusions," paper presented at the Conference on Road Charging Systems: Technology Choice and Cost Effectiveness, Paris, June 1, 2006.

Fabian, Lawrence J., "Making Cars Pay: Singapore's State-of-the-Art Congestion Management," *Transportation Planning*, Vol. 18, No. 1, 2003, pp. 1–2.

"Feuer Introduces Package of Transportation Bills to Raise Revenue for Local Projects," *California Chronicle*, February 26, 2008. As of July 28, 2008: http://www.californiachronicle.com/articles/53438

Goh, Mark, "Congestion Management and Electronic Road Pricing in Singapore," *Journal of Transport Geography*, Vol. 10, No. 1, March 2002, pp. 29–38.

Hamideh, Ashad Rafi, Jung Eun Oh, Samuel Labi, and Fred Mannering, "Econometric Analysis of Post-Election Opinion Surveys: Understanding Public Acceptance of Local Transportation Sales Taxes," paper presented at the Transportation Research Board 85th Annual Meeting, report 06-3084, Washington, D.C., January 2006.

Jeff, Gloria J., "Wilshire Boulevard Bus Lanes," memorandum to Los Angeles City Councilmember Wendy Greuel, chair, Transportation Committee, April 19, 2007.

Kumar, Ashok, Los Angeles County Metropolitan Transportation Authority, "Impact of Traffic Enforcement on Metro Rapid Running Times on Ventura Boulevard," memorandum to Ed Clifford, director of Metro Services Planning, May 2, 2007.

LADOT—*see* City of Los Angeles Department of Transportation.

Levinson, Herbert S., Samuel Zimmerman, Jennifer Clinger, Scott C. Rutherford, Rodney L. Smith, John Cracknell, and Richard Soberman, *Bus Rapid Transit*, Vol. I: *Case Studies in Bus Rapid Transit*, Washington, D.C.: National Academy Press, Transit Cooperative Research Program report 90, 2003. As of June 7, 2008: http://www.gobrt.org/CaseStudies.pdf

Los Angeles County Metropolitan Transportation Authority, *Los Angeles Metro Rapid Demonstration Program, Final Report*, Los Angeles, Calif., 2000.

———, Metro Bicycle Transportation Strategic Plan, January 2006. As of March 25, 2008:
http://www.metro.net/board/Items/2006/02_February/20060215P&PItem6%20Atta.pdf

Manville, Michael, and Donald C. Shoup, "People, Parking, and Cities," *Journal of Urban Planning and Development*, Vol. 131, No. 4, 2005, pp. 233–245.

Metro—*see* Los Angeles County Metropolitan Transportation Authority.

MITRE Corporation, U.S. Federal Highway Administration, and U.S. Department of Transportation, *Intelligent Transportation Infrastructure Benefits: Expected and Experienced*, Washington, D.C.: U.S. Department of Transportation, FHWA-JPO-96-008, January 1996. As of July 28, 2008: http://www.itsdocs.fhwa.dot.gov/jpodocs/repts_te/1mz01!.pdf

Myers, Dowell, "Promoting the Community Future in the Contest with Present Individualism," in Lewis D. Hopkins and Marisa Zapata, eds., *Engaging the Future: Forecasts, Scenarios, Plans, and Projects*, Cambridge, Mass.: Lincoln Institute of Land Policy, 2007.

Newton, Damien, "Feuer's Sales Tax and Climate Change Bills Pass Committee," *Streetsblog*, April 14, 2008. As of July 28, 2008: http://la.streetsblog.org/2008/04/14/ feuer%E2%80%99s-global-warming-and-sales-tax-bills-pass-committee/

Obenberger, Jon, "Managed Lanes," *Public Roads*, Vol. 68, No. 3, November–December 2004, pp. 48–55. As of April 23, 2008: http://www.tfhrc.gov/pubrds/04nov/08.htm

Poole, Robert W., and C. Kenneth Orski, "Hot Lanes: A Better Way to Attack Urban Highway Congestion," *Regulation*, Vol. 23, No. 1, 2000, pp. 15–20.

Pucher, John, and Lewis Dijkstra, "Making Walking and Cycling Safer: Lessons from Europe," *Transportation Quarterly*, Vol. 54, No. 3, 2000, pp. 25–50.

Rifkin, Allyn D., *Olympic/Pico One-Way Pair: Initial Feasibility Report*, Los Angeles, Calif.: Office of Los Angeles County Supervisor Zev Yaroslovsky, April 16, 2007. As of March 6, 2008: http://zev.lacounty.gov/images/OLYMPIC-PICO_REPORT-EXHIBITS_APPENDIX_4-16-07.pdf

San Diego Association of Governments, *FasTrak Value Pricing Fact Sheet*, San Diego, Calif., May 7, 2007. As of June 4, 2008: http://www.sandag.org/uploads/publicationid/publicationid_831_4185.pdf

SANDAG—*see* San Diego Association of Governments.

Santos, Georgina, and Blake Shaffer, "Preliminary Results of the London Congestion Charging Scheme," *Public Works Management and Policy*, Vol. 9, No. 2, 2004, pp. 164–181.

SCAG—*see* Southern California Association of Governments.

Schrank, David, and Tim Lomax, *The 2007 Urban Mobility Report*, College Station, Tex.: Texas Transportation Institute, Texas A&M University, September 2007. As of May 7, 2008: http://tti.tamu.edu/documents/mobility_report_2007_wappx.pdf

Short, Jack, "Conference Conclusions," paper presented at the International Conference on Managing Transport Demand Through User Charges: Experience to Date, European Conference of Ministers of Transport, London, January 23, 2004.

Shoup, Donald C., "Evaluating the Effects of Cashing Out Employer-Paid Parking: Eight Case Studies," *Transport Policy*, Vol. 4, No. 4, October 1997, pp. 201–216.

———, *The High Cost of Free Parking*, Chicago, Ill.: Planners Press, American Planning Assocation, 2005.

Sorensen, Paul A., and Brian D. Taylor, "Innovations in Road Finance: Examining the Growth in Electronic Tolling," *Public Works Management and Policy*, Vol. 11, No. 2, 2006, pp. 110–125.

Southern California Association of Governments, *Goods Movement in Southern California: The Challenge, the Opportunity, and the Solution*, Los Angeles, Calif., September 2005.

Stemley, John J., "One-Way Streets Provide Superior Safety and Convenience," *ITE Journal*, Vol. 68, No. 8, August 1998, pp. 47–50.

Sullivan, Edward C., *Continuation Study to Evaluate the Impacts of the SR 91 Value-Priced Express Lanes, Final Report*, Sacramento, Calif.: California Department of Transportation, Traffic Operations Program, HOV Systems Branch, December 2000. As of May 7, 2008:
http://ceenve3.civeng.calpoly.edu/sullivan/SR91/final_rpt/FinalRep2000.pdf

Supernak, Janusz, Jaqueline Golob, Thomas F. Golob, Christine Kaschade, Camilla Kazimi, Eric Schreffler, and Duane Steffey, "San Diego's Interstate 15 Congestion Pricing Project: Attitudinal, Behavioral, and Institutional Issues," *Transportation Research Record*, Vol. 1812, 2002, pp. 78–86.

Texas Transportation Institute, *Houston's Travel Rate Improvement Program: "Toolbox" of Improvement Strategies: Increase System Efficiency*, College Station, Tex.: Texas A&M University, April 2001. As of May 8, 2008:
http://mobility.tamu.edu/ums/trip/toolbox/increase_system_efficiency.pdf

TTI—*see* Texas Transportation Institute.

Wachs, Martin, *Improving Efficiency and Equity in Transportation Finance*, Washington, D.C.: Brookings Institution, Center on Urban and Metropolitan Policy, April 2003b. As of May 7, 2008:
http://www.brookings.edu/reports/2003/04transportation_wachs.aspx